Physics Units 3 & 4

insight.

▸ innovative ▸ engaging ▸ evolving

Copyright © Insight Publications 2024

First published in 2024
Insight Publications Pty Ltd
3/350 Charman Road
Cheltenham Victoria 3192
Australia

Tel: +61 3 8571 4950
Email: books@insightpublications.com.au

www.insightpublications.com.au

Insight VCE Revision Questions: Physics Units 3 & 4

ISBN: 9781923154995

Written and reviewed by Francis Dillon and Nick Howes
Edited by Anna Alberti
Proofread by Geoffrey Marnell
Cover design and internal layout by Melisa Paredes
Internal design by Bec Yule @ Red Chilli Design
Images by Oscar Puentes
Printed by Markono Print Media Pte Ltd

Insight Publications acknowledges the Traditional Custodians of the Country on which we meet and work, the Boonwurrung People of the Kulin Nation. We pay our respects to their Elders past and present, and extend that respect to all Aboriginal and Torres Strait Islander Peoples.

● Contents

● Introduction

Insight's *VCE Revision Questions: Physics Units 3 & 4* contains questions, worked solutions and tips to help you develop skills for assessment. The questions cover all areas of study in Units 3 and 4 of VCE Physics. A good habit to implement is to test yourself by working through this resource. The process of actively recalling information assists with deeper learning, and you will be able to compare your answers with the provided worked solutions.

By using this resource as part of your study regimen throughout the year, you will be prepared for questions you may encounter in your end-of-year VCE exam.

We wish you well with your studies.

The Insight Team

🔵 Questions

Unit **3** | Area of Study **1** How do physicists explain motion in two dimensions?

Note: You can find the formula sheet on the VCAA (Victorian Curriculum and Assessment Authority) webpage for the end-of-year Physics assessment, https://www.vcaa.vic.edu.au/assessment/vce-assessment/past-examinations/Pages/physics.aspx .

SET 1

Use the following information to answer Questions 1 and 2.

A pendulum is allowed to swing freely, as shown in the diagram below.

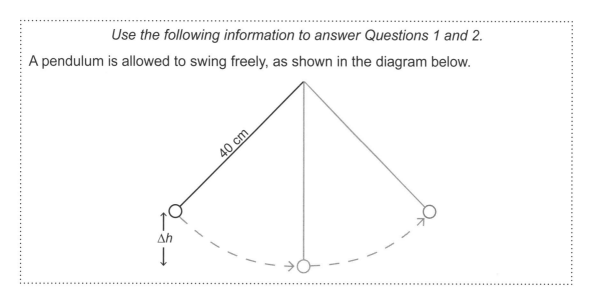

Question 1

Which one of the following graphs best shows how kinetic energy (E_k) changes with height (Δh)?

Take the bottom of the swing as $h = 0$ and ignore friction.

A.

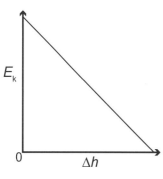

B.

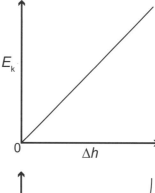

C.

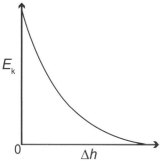

D.

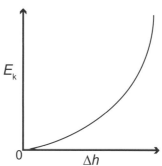

Question 2

At the bottom of its motion, the pendulum (mass 60 g) is found to be moving at 2.3 m s^{-1}.

Which one of the following best describes the tension in the string at this time?

A.   0.6 N

B.   1.4 N

C.   2.0 N

D.   2.6 N

Question 3

A passenger airliner, with a total mass of 15 000 kg, is making a circular horizontal turn at a constant speed, as shown in the diagram below. The wings are experiencing a net force, directed towards the centre of the circle, of 125 500 N. The circular horizontal turn has a radius of 690 m.

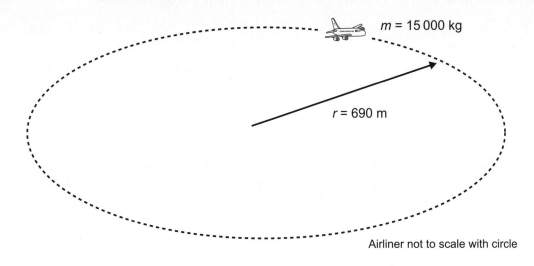

m = 15 000 kg

r = 690 m

Airliner not to scale with circle

Which of the following is closest to the speed of the airliner?

A.   57.8 m s^{-1}

B.   72.1 m s^{-1}

C.   76.0 m s^{-1}

D.   13.4 m s^{-1}

SET 2

> *Use the following information to answer Questions 1 and 2.*
>
> Erica, who has a mass of 55 kg, is seated in the Rocketing Upwards ride, which accelerates her initially upwards at a rate of 3.0 m s^{-2}.

Question 1

Which one of the following is the correct magnitude and direction of the force that Erica exerts on the seat of the ride?

A.   374 N upwards

B.   374 N downwards

C.   704 N upwards

D.   704 N downwards

Question 2

Erica travels from **rest** at ground level ($h = 0$ m), then **stops** at the highest level where $h = 20$ m.

Which one of the following is closest to the work done on her by the ride?

A.   1078 J

B.   10 780 J

C.   20 780 J

D.   1100 J

Question 3

Which one of the following distance–time graphs represents a moving object that is slowing down to a stop?

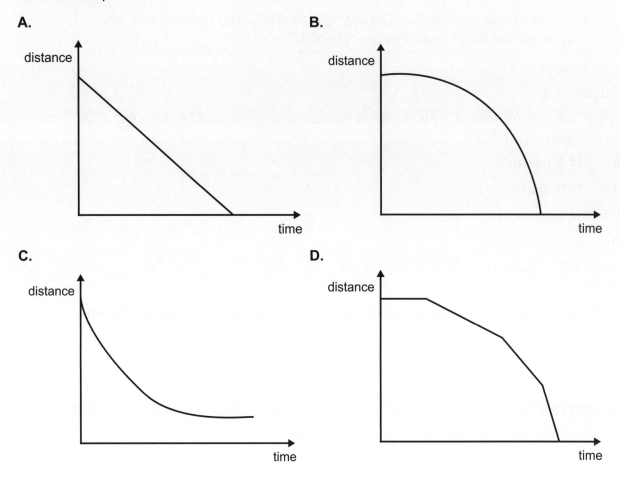

A.

distance

time

B.

distance

time

C.

distance

time

D.

distance

time

Question 4

A circular ring is pulled by three forces, F_A = 12 N, F_B = 13 N and F_C = 5 N, as shown below. The net force on the ring is 0 N. The effects of the size of the ring and friction may be considered negligible.

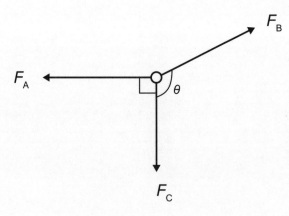

F_B

F_A

θ

F_C

Which of the following is closest to the size of θ, the angle between F_B and F_C?

A. 113°

B. 115°

C. 155°

D. 157°

SET 3

Use the following information to answer Questions 1 and 2.

A 12 kg ball is attached to a length of rope with a breaking strength of 750 N. The ball travels in uniform horizontal circular motion with a radius of 2.3 m, as shown below.

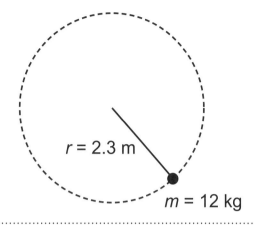

$r = 2.3$ m

$m = 12$ kg

Question 1

The maximum speed of the ball that will not break the rope is closest to

A. 1.2 m s^{-1}

B. 6.3 m s^{-1}

C. 12 m s^{-1}

D. 63 m s^{-1}

Question 2

The ball is already travelling at the maximum speed allowed by the breaking strength of the rope.

If the speed of the ball is maintained at this value, which one of the following changes will cause the rope to break?

A.   increase the radius of the uniform circular motion

B.   decrease the radius of the uniform circular motion

C.   increase the breaking strength of the rope

D.   decrease the mass of the ball

Question 3

A mass, m, is attached to a spring of natural length, l. The mass is allowed to fall under gravity and the spring reaches maximum extension, as shown below. Air resistance may be ignored.

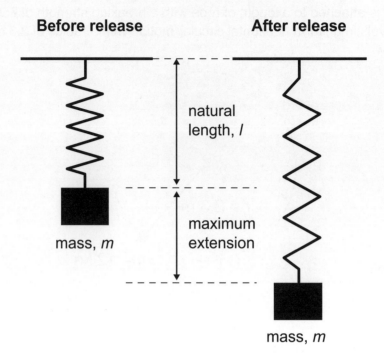

Which one of the following statements is correct?

A. At maximum extension, the total energy of the mass is the same as that at zero extension.

B. At maximum extension, the kinetic energy of the system is maximum.

C. At zero extension, the total energy of the mass is maximum.

D. At zero extension, the kinetic energy of the system is maximum.

Question 4

An aeroplane is flying at a constant speed in a horizontal circle.

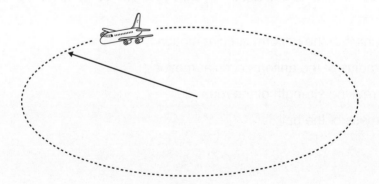

Which one of the following statements best describes the motion of the aeroplane?

A. The net force on the aeroplane is zero because it is moving at a constant speed.

B. The velocity of the aeroplane is constant because it is flying in a horizontal circle.

C. The thrust of the aeroplane's engines is greater than the air resistance, which provides centripetal acceleration.

D. The upward lift of the aeroplane's wings is equal to the force due to gravity on the aeroplane.

SET 4

Question 1

Taku and Mina are experimenting with a frictionless air track. A cart (mass M = 480 g) is attached to a falling mass (m = 80 g), using a massless cable. The falling mass is allowed to free fall under the influence of gravity. The cable runs over a frictionless pulley and air resistance is negligible.

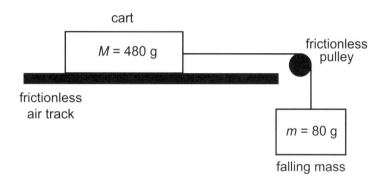

Taku and Mina measure the acceleration of the cart using an onboard accelerometer, the mass of which is included in the mass of the cart.

Which one of the following is closest to the expected magnitude of the acceleration of the cart?

A. 1.2 m s^{-2}

B. 1.4 m s^{-2}

C. 1.6 m s^{-2}

D. 1.8 m s^{-2}

Question 2

A group of physics students, Ali, Bobbi, Curtis and Daria, are waiting at a bus stop. They observe a group of passengers boarding a bus that is heading to a different destination. As the bus accelerates away from the bus stop, the students notice a few standing passengers appearing to fall towards the rear of the bus. They discuss their observations.

Which one of the students best describes the physics of the motion of the standing passengers?

A. Ali   'The force of inertia on these passengers caused them to fall towards the rear of the bus.'

B. Bobbi   'There is an imaginary force acting on these passengers, making them fall towards the rear of the bus.'

C. Curtis   'The bus exerts a backward force on the passengers to make them fall towards the rear of the bus.'

D. Daria   'Friction force on the passengers' feet pulls their legs forward, but their bodies fall towards the rear of the bus.'

SET 5

Question 1 (5 marks)

Two boxes, labelled A and B, are stacked one on top of the other in a lift, as shown below. The mass of box A is 3 kg and the mass of box B is 7 kg. The lift is currently stationary.

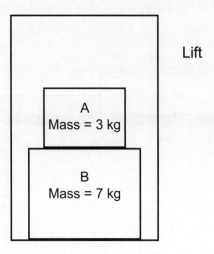

Lift

A
Mass = 3 kg

B
Mass = 7 kg

a.   Calculate the magnitude of the force exerted on box A by box B, and state the direction of the force, using either *up* or *down*.   2 marks

N		direction:

b.   The lift now accelerates down at $a = 3.5$ m s^{-2}.

Calculate the magnitude of the force on box B by box A, and state the direction of the force, using either *up* or *down*.   3 marks

N		direction:

Question 2 (11 marks)

Loadzafun Park has a ride consisting of a carriage, C1, starting at point P (as shown below), rolling down a set of smooth, frictionless rails and then colliding with a stationary carriage, C2, at point Q.

The total mass of carriage C1, including passengers, is 850 kg, and its initial velocity is 3.5 m s^{-1}.

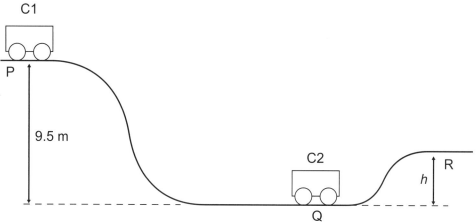

a. What is the velocity of carriage C1 just before it collides with carriage C2 at point Q?

3 marks

	m s^{-1}

The total mass of carriage C2 is 550 kg, and it is stationary prior to carriage C1 colliding with it. The collision may be considered an isolated collision. After the collision, both carriages stick together and travel up the rail to point R, where they both come to a halt.

b. Explain what is meant by 'isolated collision', stating the necessary condition for it to occur and the outcome for the momentum of the system.   3 marks

_____

_____

_____

_____

c. Determine the velocity of the combined carriages just after the collision.   2 marks

_____

_____

_____

m s⁻¹

d. Calculate the height, h, of point R above the level of Q.   3 marks

_____

_____

_____

_____

m

Question 3 (5 marks)

A parcel with a mass of 4.5 kg is placed on the roof of a 750 kg car that is travelling at a speed of 17 m s^{-1} and going around a roundabout with a radius of 8 m. The parcel remains motionless relative to the roof of the car due to the friction between the parcel and the roof; however, it is about to begin to slide.

a. Calculate the magnitude of the friction force of the roof of the car on the parcel.

2 marks

	N

b. The car now speeds up to 19 m s^{-1}.

What is the minimum radius that the car should travel around the roundabout so that the parcel remains motionless relative to the roof of the car?

3 marks

	m

Question 4 (3 marks)

Teng and Sophie are discussing an experiment involving a mass suspended on a spring, as shown in the diagram below. Initially, the position of the mass is at the natural length of the spring. The mass is released and allowed to fall freely under gravity and to extend the spring. At the maximum extension of the spring, the mass stops momentarily prior to it springing back upwards.

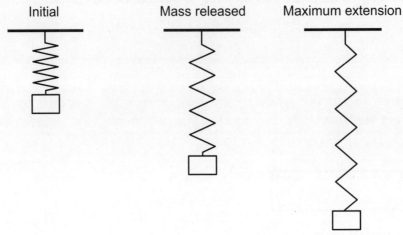

Initial   Mass released   Maximum extension

Regarding the kinetic energy of the mass as it falls between these two positions, Teng and Sophie make the following statements.

Teng says, 'At the halfway point where the force due to gravity on the mass equals the elastic force of the spring, the kinetic energy of the system reaches its maximum.'

Sophie says, 'At the halfway point, the majority of the total energy of the system is in the form of kinetic energy.'

Evaluate their opinions. Detailed calculations are **not** necessary.

SET 6

Question 1 (8 marks)

During crash testing, a 200 kg block is dropped and allowed to swing into an 85 kg crash test dummy, as shown below.

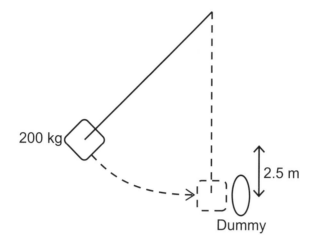

a. If the block falls 2.5 m, how much kinetic energy is delivered to the dummy, assuming all of the block's energy is transferred?   2 marks

b. During another crash test, the 200 kg block is dropped from 7.2 m and hits the dummy at 20 m s^{-1}. After the collision, the block and the dummy move off together.

Calculate their velocity immediately after the collision.   2 marks

m s^{-1}

c. Show that this collision is inelastic.   2 marks

_____

_____

_____

d. In another crash test, the 200 kg block is used to break a piece of wood, as shown below.

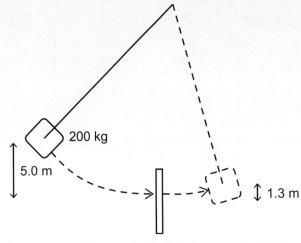

If the block is dropped from 5.0 m, breaks the piece of wood and continues to
a height of 1.3 m, how much energy was used to break the wood?   2 marks

_____

_____

_____

	J

Question 2 (3 marks)

A 63 g tennis ball is attached to a light string and swung in a circle, as shown below.

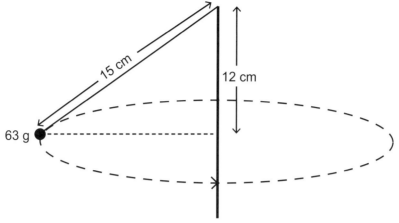

Use the information provided to calculate the net force acting on the ball. Clearly show each step of your working.

	N

Question 3 (5 marks)

David stands at the edge of a 6.0 m tall building and wants to kick his soccer ball onto the roof of the building across the road, as shown in the diagram immediately below.

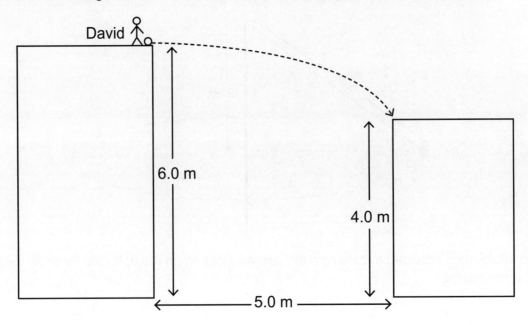

a.   If the ball is kicked horizontally, what is the minimum initial velocity required to land the ball on the opposite roof?   3 marks

_____

_____

_____

_____

_____

_____

_____

_____

_____

m s^{-1}

As shown in the diagram immediately below, David leans over the edge of his building and throws the ball straight up at 4.9 m s^{-1}.

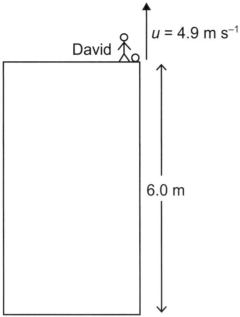

David

$u = 4.9$ m s^{-1}

6.0 m

b. How fast is the ball moving when it hits the ground? 2 marks

	m s^{-1}

Unit 3 | Area of Study 2   How do things move without contact?

SET 1

Question 1

Which one of the following diagrams best shows the magnetic field pattern between two south poles?

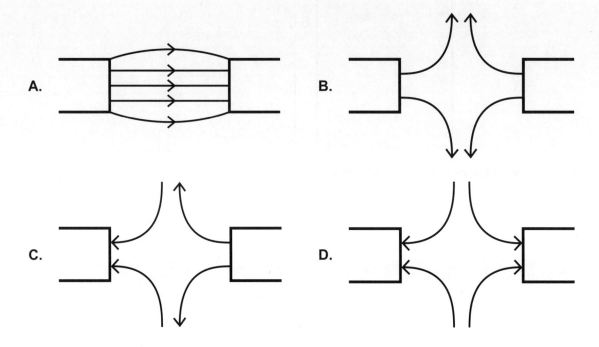

Question 2

A wire carrying a current into the page is placed in a magnetic field, as shown in the diagram below.

Which one of the following describes the direction of the force on the wire caused by the magnetic field?

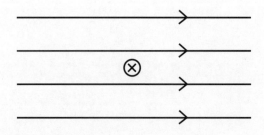

A. up the page

B. down the page

C. left

D. right

Question 3

Two small spheres hold electric charges of +0.01 mC and +0.03 mC and are placed 60 cm apart.

Which one of the following best describes the magnitude and direction of the force on each sphere?

Use $k = 8.99 \times 10^9$ N m^2 C^{-2}.

A.   7.5 N repulsion

B.   7.5 N attraction

C.   7.5 MN repulsion

D.   7.5 MN attraction

Question 4

An electron is accelerated between two plates through a potential difference of 5000 V.

How much kinetic energy does the electron gain from the electric field between these plates? Ignore relativistic effects.

A. 8.00×10^{-16} J

B. 5.00×10^{-16} J

C. 4.20×10^7 J

D. 1.70×10^{15} J

Question 5

The three diagrams below, X, Y and Z, represent fields.

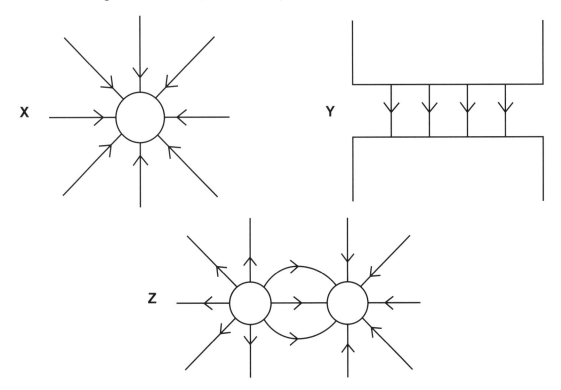

Which one of the following could correctly identify the field types?

	X	Y	Z
A.	electric	magnetic	gravitational
B.	gravitational	electric	gravitational
C.	magnetic	gravitational	electric
D.	gravitational	magnetic	electric

Use the following information to answer Questions 6 and 7.

An electron gun ejects electrons horizontally at a constant velocity into a vacuum chamber. The electrons are deflected with a constant vertical force downwards, so that they hit a target that is 39 cm below the horizontal path of the electrons. The flight time of the electrons is 2.3×10^{-4} s.

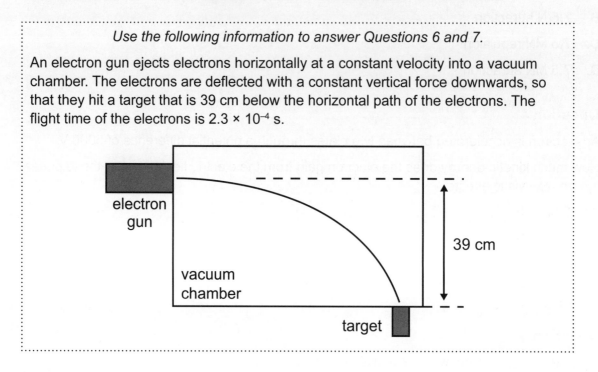

Question 6

Which one of the following is the closest value of the **downward** acceleration of the electrons?

A. 1.5×10^7 m s^{-2}

B. 1.5×10^8 m s^{-2}

C. 1.5×10^9 m s^{-2}

D. 1.5×10^{10} m s^{-2}

Question 7

The constant vertical acceleration of the electrons may be caused by

A.   gravity only.

B.   an electric field only.

C.   a magnetic field only.

D.   either an electric field or a magnetic field.

SET 2

Question 1

An electron travelling in a straight line enters a field. The force exerted by the field is perpendicular to the electron's original path.

Which one of the following types of fields could achieve this deflection?

A. a gravitational field only

B. either an electric field or a magnetic field

C. an electric field only

D. either an electric field, a gravitational field or a magnetic field

Question 2

Duke is orbiting above the planet Zorb, which has a radius of r, as shown in the diagram below. He measures the gravitational field strength at an altitude of r above the surface of the planet to be $g = 9$ N kg^{-1}.

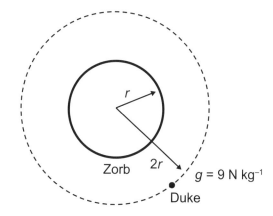

When Duke's altitude increases to $2r$ above the surface, what will be the gravitational field strength at this new position?

A. $g = 4.5$ N kg^{-1}

B. $g = 2.25$ N kg^{-1}

C. $g = 4$ N kg^{-1}

D. $g = 1$ N kg^{-1}

Question 3

A metallic object is brought close to the north pole of a bar magnet. The magnetic field of the magnet exerts

A. an attractive force only.

B. either an attractive force or a repulsive force.

C. a repulsive force only.

D. either no force, an attractive force or a repulsive force.

Question 4

Which one of the following fields may be described as static and non-uniform?

A. the gravitational field around a point mass

B. the electric field between two parallel plates at a constant potential difference

C. the magnetic field around a solenoid connected to an AC power supply

D. the electric field between two plates connected to an AC power supply

Question 5

The magnetic field lines around the Earth run from the South Pole to the North Pole, as shown in the diagram below. A current of 1.5 A flows through a wire on the surface of the Earth from east to west.

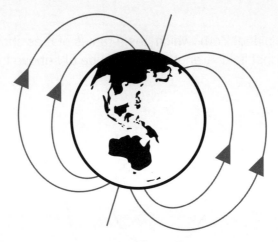

The wire will experience a force in the direction of

A. up.

B. down.

C. north.

D. south.

SET 3

Question 1

Consider two positively charged particles travelling at a constant speed, v. One particle enters a uniform electric field, with field strength E, between two parallel plates, as shown below (left). The other particle enters a uniform magnetic field, with field strength B, directed out of the page, as shown below (right).

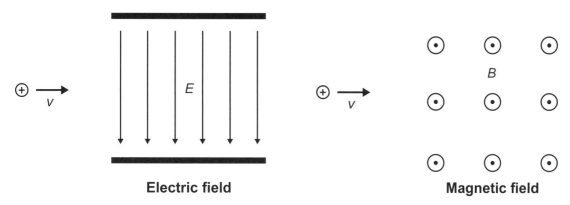

What is the shape of the path of each particle in each of the force fields?

	Electric field	Magnetic field
A.	parabolic	parabolic
B.	parabolic	circular
C.	circular	circular
D.	circular	parabolic

Use the following information to answer Questions 2 and 3.

An alpha particle, which has a +2 charge, is stationary in deep space.

Question 2

Determine the magnitude of the force that the alpha particle exerts on an electron that is brought to a position 5 μm from it.

A. 115 N

B. 1.84×10^{-17} N

C. 1.84×10^{-23} N

D. 9.21×10^{-23} N

Question 3

The electron is now removed, leaving the alpha particle on its own.

Which one of the following describes the electric field around the alpha particle?

A. changes with time

B. uniform

C. non-uniform

D. moving

Question 4

A bundle of 10 wires, each carrying a current of 0.70 A, is placed in a uniform magnetic field. A straight section of the bundle, with a length of 0.45 m, is entirely within the magnetic field. The bundle of wire experiences a force of 3.7 mN.

What is the strength of the magnetic field?

A. 1.17×10^{-1} T

B. 1.17×10^{-2} T

C. 1.17×10^{-3} T

D. 1.17×10^{-4} T

Question 5

The strength of the gravitational field reduces with distance from Earth's centre of mass.

What is the gravitational potential, to 3 significant figures, at the peak of Mount Kosciuszko, which is 6.38×10^6 m from Earth's centre of mass?

A.   9.79 N kg^{-1}

B.   9.78 N kg^{-1}

C.   9.81 N kg^{-1}

D.   9.82 N kg^{-1}

SET 4

Question 1

Both the Kosmos-2251 satellite and the Iridium 33 satellite orbit Earth at the same altitude.

Which one of the following statements must be correct regarding the motion of both satellites?

A.   The mass of Kosmos-2251 is the same as the mass of Iridium 33.

B.   The speed of Kosmos-2251 is the same as the speed of Iridium 33.

C.   Both Kosmos-2251 and Iridium 33 orbit Earth along the same orbital path.

D.   The relative speeds of the satellites are directly proportional to their relative masses.

Question 2

An electrical field is generated in the space around a point electric charge or between a pair of plates connected to a DC supply.

Which one of the following statements describes both types of electric field?

A.   Their field strength follows the inverse square law.

B.   Their field lines point away from the positive charge or plate.

C.   Their field strength is uniform.

D.   Their field lines point towards the positive charge or plate.

Question 3

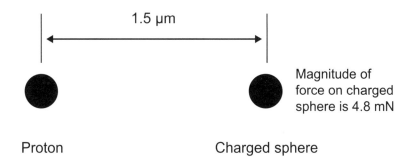

1.5 μm

Magnitude of force on charged sphere is 4.8 mN

Proton   Charged sphere

A tiny charged sphere is experiencing a force of 4.8 mN when placed at a distance of 1.5 μm from a proton (q_{proton} = 1.6 × 10⁻¹⁹ C).

What is the amount of charge carried by the tiny sphere?

A. 7.5×10^{-4} C

B. 7.5×10^{-5} C

C. 7.5×10^{-6} C

D. 7.5×10^{-7} C

Question 4

P   ●

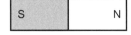   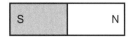

Two identical bar magnets are placed in line with each other. Point P is equidistant from the north pole of the magnet on the left, and from the south pole of the magnet on the right.

Which one of the following arrows best represents the direction of the magnetic field at point P?

A.   →

B.   ←

C.   ↘

D.   ↑

Question 5

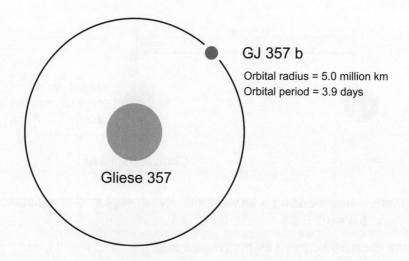

GJ 357 b

Orbital radius = 5.0 million km
Orbital period = 3.9 days

Gliese 357 is a star in the Hydra constellation. It has been confirmed that there are three planets orbiting Gliese. One planet, known as GJ 357 b, orbits at a radius of 5.0 million km with a period of 3.9 days. Another planet, which is known as GJ 357 d, orbits with a period of 56 days.

Which one of the following is closest to the orbital radius of GJ 357 d?

A. 27 million km

B. 30 million km

C. 44 million km

D. 72 million km

SET 5

Question 1 (3 marks)

The diagram below shows a single point electric charge of +2.0 µC at point A. At point B is a single point charge of +e.

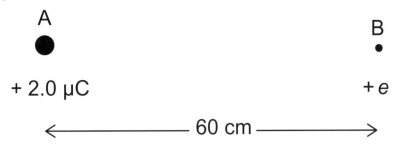

a.   Use an arrow to show the direction of the force on B due to A.   1 mark

b.   Calculate the strength of the electric field due to A at point B.

Use $k = 8.99 \times 10^9$ N m² C⁻².   2 marks

	N C⁻¹

Question 2 (5 marks)

An electron moving at 1.5×10^7 m s^{-1} to the right passes through a magnetic field of strength 8.0 mT into the page, as shown below.

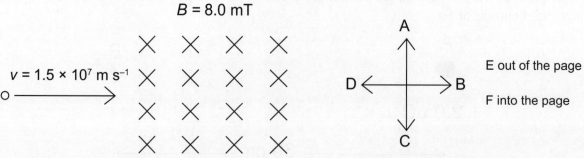

a.   Calculate the magnitude of the force that acts on the electron as it moves through the field.   2 marks

N

b.   Using the direction guide A–F shown in the diagram above, which direction best shows the direction of the force on the electron as it enters the field?   1 mark

c.   Calculate the radius of the electron's path through the magnetic field.   2 marks

m

Question 3 (6 marks)

In a distant solar system, a planet is observed orbiting a star. The mass of the star is estimated to be 2.9×10^{30} kg and the planet is believed to have an orbital radius of 1.4×10^{12} m about the star, as shown in the diagram below.

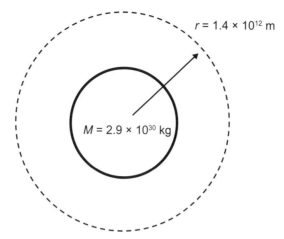

$r = 1.4 \times 10^{12}$ m

$M = 2.9 \times 10^{30}$ kg

a. Calculate the gravitational field strength of the star at the orbital radius of the planet.

2 marks

N kg^{-1}

b. Determine the acceleration of the planet about the star.

1 mark

m s^{-2}

c. Is it possible to determine the mass of the planet from the information given
    in this question? Give an explanation for your answer.   3 marks

_____

_____

_____

_____

_____

_____

_____

SET 6

Question 1 (4 marks)

Two charges are arranged in a vertical line, as shown in the diagram below. Point X is midway
between the two charges. The charges are fixed in the positions shown and are unable to move.

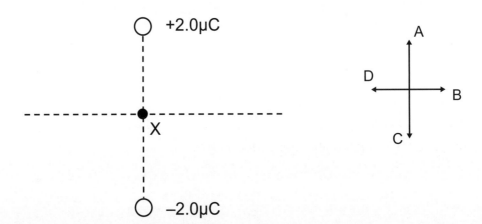

a. On the diagram above, draw an arrow to represent the direction of the electric
    field strength at the point X due to both charges.   1 mark

b. A small negatively charged sphere is now positioned at X and released.

    State the direction (using the direction guide A–D given in the diagram) in which
    the sphere would move after it is released.   1 mark

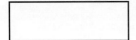

c. The distance between the two charges is 0.50 m.

Calculate the magnitude of the force acting on each charge due to the presence of the other charge. 2 marks

	N

Question 2 (3 marks)

The diagram below shows a schematic diagram of a conductor that is completely within a uniform magnetic field of 0.8 T. The length of the conductor is 12 cm. The magnitude of the current through the conductor is 0.6 A, in the direction shown on the diagram.

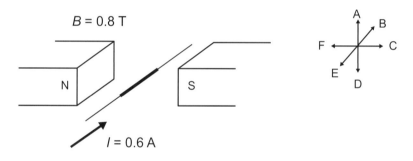

Determine the size and direction (using the direction guide A–F) of the force acting on the conductor.

	N		direction:

Question 3 (5 marks)

The diagram below shows a DC motor consisting of multiple loops of wire in a uniform magnetic field.

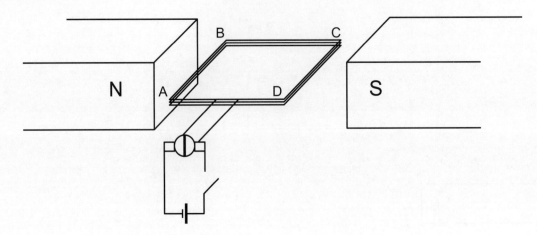

a. When the switch is closed, current flows through the loops of wire.

Circle **one** of the options below to indicate in which direction the loops rotate, as viewed from the commutator end, and explain your answer.   3 marks

clockwise / anticlockwise / no movement

_____

_____

_____

_____

b. Describe what happens to the current in the loop as the coil rotates past the vertical position. Explain why this needs to occur.   2 marks

_____

_____

_____

SET 7

Question 1 (6 marks)

In 2015, the NASA probe *New Horizons* made the closest ever fly-by of the dwarf planet Pluto, coming within 12 500 km of Pluto's surface.

Mass of probe	410 kg
Mass of Pluto	1.3×10^{22} kg
Radius of Pluto	1200 km
Gravitational constant, G	6.67×10^{-11} N m² kg⁻²

a. Calculate the strength of the gravitational attraction between Pluto and the *New Horizons* probe at an altitude of 12 500 km. Use the most appropriate number of significant figures in your answer. **3 marks**

	N

The graph below shows the gravitational field strength above the surface of Pluto.

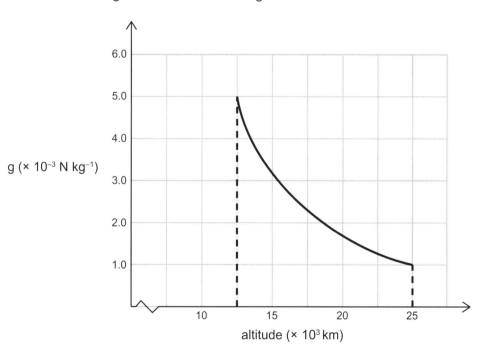

b. Use the graph on the previous page to calculate the kinetic energy gained by the probe as it moves from an altitude of 25 000 km to 12 500 km. Clearly show all of your working.

3 marks

J

Question 2 (8 marks)

Electric and magnetic fields can be used to change the speed and direction of motion of charged particles. The diagram below shows a schematic diagram of an electron gun. The plates X and Y are used to accelerate electrons emitted from the tungsten filament, and in region Z a uniform magnetic field is used to direct the beam onto a particular point on the target.

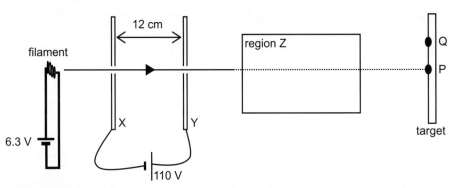

accelerating voltage = 110 V

mass of electron = 9.11×10^{-31} kg

charge of electron = 1.60×10^{-19} C

a. Calculate the magnitude of the electric field strength between the plates X and Y.

2 marks

V m^{-1}

b. The electrons emitted from the filament can be assumed to be initially at rest as they pass through plate X.

Determine the speed of the electrons entering region Z after being accelerated by the electric field.   3 marks

	m s^{-1}

c. When the magnetic field in region Z is zero, the electrons strike the target at P. The beam of electrons is required to be moved to point Q by altering the magnetic field strength in region Z.

Using your knowledge of magnetic fields and force, explain in which direction the magnetic field in region Z must be so that the beam of electrons strikes the target at Q. When describing your chosen direction, use one of *up, down, left, right, into the page* or *out of the page*.   3 marks

SET 8

Question 1 (5 marks)

A metal sphere, labelled S1, carries 8.0 μC of positive charge. The radius of the sphere is not significant so S1 may be considered as a point charge. A second similar sphere, labelled S2, but without any charge, is placed at a distance d = 5.8 cm from S1, as shown below.

Both spheres are held in place with perfectly insulating rods.

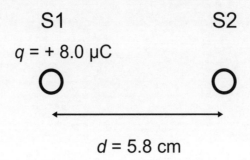

a. State the magnitude of the electrical force exerted by S1 on S2.   1 mark

N

b. Sphere S2 is then brought into contact with sphere S1. After a few seconds, it is returned to its original position.

What is the magnitude of the charge on S2 now? Give your answer correct to 2 significant figures.   2 marks

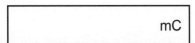
mC

c. On the diagram above, draw an arrow from the centre of S1 to indicate the direction of the force that S2 will exert on it. Explain your answer.   2 marks

Question 2 (6 marks)

A stationary loop of metal wire with a cross-sectional area of 13.7 cm² is placed completely inside the magnetic field of an electromagnet, as shown in the diagram immediately below. The loop is connected to a voltmeter, which measures any EMF generated between the two terminals of the loop.

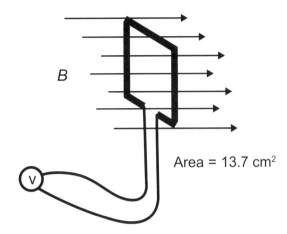

Area = 13.7 cm²

The electromagnet is initially switched off. The current that is used to generate the electromagnet is then varied such that the magnetic field strength changes according to the graph, as shown below.

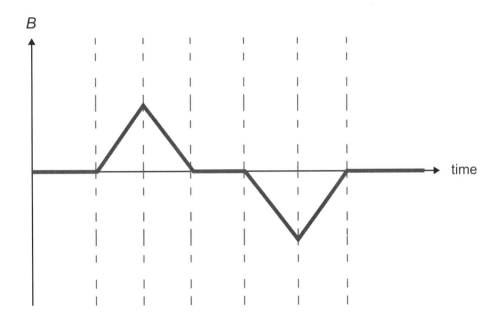

a. Sketch the EMF induced in the loop as the magnetic field strength of the electromagnet varies. You do not need to include any values on the axes.   3 marks

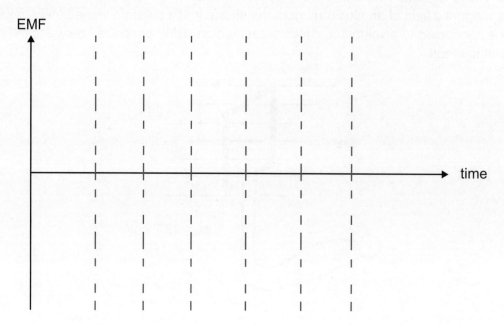

The electromagnet is switched off. It is then switched on and the magnetic field strength increases linearly to a value of $B = 0.06$ T in a time of 0.09 s.

b. What is the maximum magnetic flux through the loop?   1 mark

Wb

c. Calculate the magnitude of the EMF generated in the loop in this time period. Show your working.   2 marks

V

Question 3 (4 marks)

A U-shaped magnet is used to study the effect of a magnetic field on a conductor carrying a current, as shown below. The conductor receives current from a pair of wires that are wholly outside the magnetic field of the magnet.

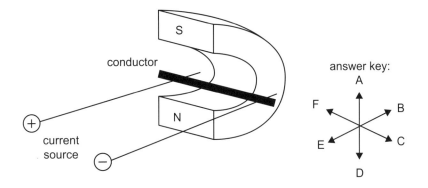

a. The current source is switched on and the conductor experiences a force due to the magnetic field of the magnet. Use the answer key to state the direction of the force on the conductor. Explain your answer. 2 marks

b. The length of the conductor that is within the magnetic field is 6.5 cm. The current through the conductor is 0.39 A. A force meter measures the magnetic force on the conductor as 7.3×10^{-6} N.

Determine the strength of the magnetic field of the U-shaped magnet. Show your working. 2 marks

T

SET 9

Question 1 (5 marks)

The OSIRIS-REx is a space probe sent to the asteroid 101955 Bennu to collect a sample and return it to Earth. The probe reached the region of the asteroid in December 2018. For the next 2 years it orbited the asteroid in order to study it from afar and to calculate its mass accurately in preparation for a safe landing.

After making several hundred orbits around the asteroid in a month, the orbital data of the probe was found to be

- orbital diameter: 3480 m
- orbital period: 209 000 s.

a.   Using the orbital data of OSIRIS-REx provided above, calculate the mass of asteroid 101955 Bennu.   3 marks

	kg

b.   101955 Bennu may be considered as a sphere with a radius of 300 m. When OSIRIS-REx landed on its surface, what was the expected gravitational field strength on its surface?   2 marks

	N kg^{-1}

Question 2 (4 marks)

A 9 V DC supply is used to heat up a tungsten filament to generate free electrons in a vacuum chamber. The electrons may be considered to be stationary at this point. They are then accelerated between two charged plates that are at a potential difference of 2.4 MV DC, as shown below.

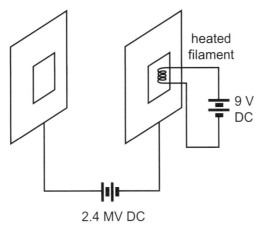

2.4 MV DC

a. Using the classical physics equations for electrical potential energy and kinetic energy, show that the expected velocity of the electrons when they arrive at the positively charged plate is equal to 9.19×10^8 m s^{-1}.   2 marks

b. Explain why the expected velocity of the electrons cannot be attained when calculated using classical physics.   2 marks

Question 3 (6 marks)

India is among the most active nations in launching satellites to fulfil tasks in telecommunications, navigation and meteorology. In 2019 it launched RISAT-2BR1 to an altitude of 576 km above Earth's surface.

Note that the mass of Earth is 5.97×10^{24} kg and its radius is 6.37×10^6 m.

a. Determine the orbital period for RISAT-2BR1.   4 marks

	s

b. India plans to launch more satellites to the same altitude as RISAT-2BR1. The mass of each of these satellites will be approximately 25% larger than that of RISAT-2BR1.

From the options below, circle all of the characteristics of the new satellites that will be the same as those of RISAT-2BR1.   2 marks

centripetal acceleration   centripetal force

orbital period   orbital speed

SET 10

Question 1 (7 marks)

A GPS satellite completes a circular orbit of Earth at a radius of 27×10^6 m, as shown below.

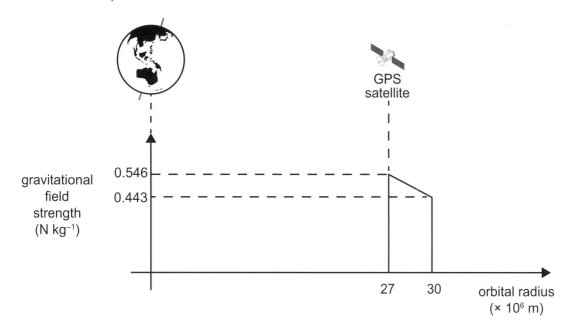

a. Show that the gravitational field strength of Earth is 0.546 N kg^{-1} at this orbital radius.   2 marks

_____

_____

_____

_____

b. Calculate the orbital speed of the GPS satellite at this orbital radius.   2 marks

_____

_____

_____

_____

m s^{-1}

The GPS satellite is boosted to a higher altitude, to an orbital radius of 30×10^6 m. At this new position, the gravitational field strength of Earth is 0.443 N kg^{-1}. The variation in the gravitational field strength may be assumed to be linear, as shown in the diagram on the previous page. The mass of the GPS satellite is 1630 kg.

c.   Using the graph shown in the diagram on the previous page, calculate the difference in the gravitational potential energy of the satellite between these two positions.   3 marks

J

Question 2 (8 marks)

A model electric motor has a rectangular coil made with 20 loops of copper wire. The length of side KL is 2.2 cm, as shown in the figure below. The loop is entirely within a magnetic field of 4.5 mT and connected to a split-ring commutator with terminals labelled P and Q. P is connected to the positive terminal of a 9 V DC supply, and Q is connected to the negative terminal. A student measures the resistance of the circuit as $R = 18\ \Omega$.

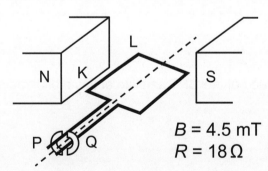

$B = 4.5$ mT
$R = 18\,\Omega$

a.   i.   Circle the direction the motor will turn when viewed from the commutator side.   1 mark

        clockwise   anticlockwise   the motor will not turn

ii. Explain your answer to **part a.i.** 2 marks

b. Calculate the magnitude of the magnetic force on side KL. 3 marks

```
┌─────────────────────────────┐
│                        N    │
└─────────────────────────────┘
```

c. A student decides to switch the power supply to 9 V AC while the rectangular coil is in the position shown in the diagram on the previous page.

Explain what will happen to the motion of the rectangular coil. 2 marks

SET 11

Question 1 (13 marks)

An electron travels horizontally at a constant velocity of 3.5×10^6 m s^{-1} into a uniform electric field between two parallel plates that are connected to a 100 V DC supply. The distance between the two plates is $d = 40$ cm. Plate A is connected to the negative terminal of the DC supply, and plate B is connected to the positive terminal of the DC supply. The motion of the electron as it enters and travels through the uniform electric field is shown in the diagram below. Ignore the effects of gravity, air resistance or special relativity.

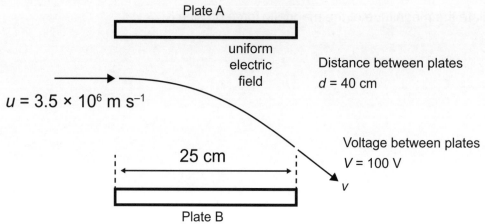

Plate A

uniform electric field

$u = 3.5 \times 10^6$ m s^{-1}

25 cm

Distance between plates
$d = 40$ cm

Voltage between plates
$V = 100$ V

v

Plate B

a. Explain why the path of the electron is the shape of a parabola while the electron is inside the uniform electric field. 3 marks

b. Show that the field strength of the uniform electric field between the plates is 250 V m^{-1}. Identify an alternative unit to V m^{-1}. 3 marks

Alternative unit:

c. Calculate the acceleration of the electron while it is inside the uniform electric field.   3 marks

$$\boxed{} \text{ m s}^{-2}$$

d. Calculate the speed of the electron, v, as it exits the uniform electric field.   4 marks

$$\boxed{} \text{ m s}^{-1}$$

Unit 3 | Area of Study 3   How are fields used in electricity generation?

SET 1

Question 1

Which one of the following actions would **not** increase the EMF generated by a rotating coil inside a magnetic field?

A.   increasing the period of the rotating coil

B.   increasing the strength of the magnetic field

C.   increasing the number of loops in the rotating coil

D.   increasing the speed of rotation of the coil

Question 2

Power transmission is commonly carried out by stepping up the voltage before transmission. After transmission, the voltage is stepped down prior to reaching consumer households.

Which one of the following statements is the **most correct** explanation of why transformers are used in the transmission of power through power lines?

A.   Stepping up and stepping down the voltage results in zero loss of power in the transmission line.

B.   Stepping up the voltage increases the power by the factor of the turns ratio, resulting in more power being delivered to the transmission lines.

C.   The process of stepping up the voltage of the electricity before transmission reduces the amount of current present in the lines, and hence reduces the amount of power lost in the lines.

D.   Conservation of energy means the resistance of the transmission line is reduced and this is the key factor in lowering the transmission losses.

Use the following information to answer Questions 3 and 4.

A mobile phone charger comprises a step-down transformer and a rectifier (which converts the AC output of the transformer to DC). The charger is connected to a 240 V_{RMS} AC power outlet, and the output of the transformer is 15.0 V_{RMS} AC.

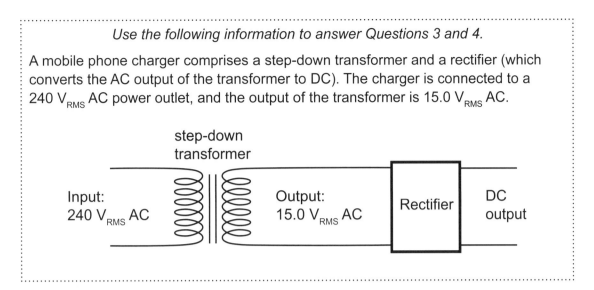

Question 3

What is the turns ratio of the step-down transformer?

A. 1:16

B. 15:1

C. 16:1

D. 24:1

Question 4

Assuming that the transformer and the rectifier are ideal, the DC output of the mobile phone charger is expected to be

A. 10.6 V DC

B. 15.0 V DC

C. 21.2 V DC

D. 30.0 V DC

Question 5

An AC generator uses a coil of 50 turns to produce 110 V_{RMS} AC at a frequency of 60 Hz.

Which one of the following actions will double the magnitude of the output voltage but has no other effect on the output voltage wave?

A. rotating the coil twice as fast

B. increasing the number of turns to 100

C. halving the period of rotation

D. increasing the magnetic field strength by 200%

Question 6

A group of students is generating electricity using a model generator shown in the diagram below on the left. The model generator uses a rotating rectangular coil that is connected to a pair of slip rings. The output voltage between the terminals labelled X and Y is shown on the graph. The slip rings are now replaced with a split-ring commutator.

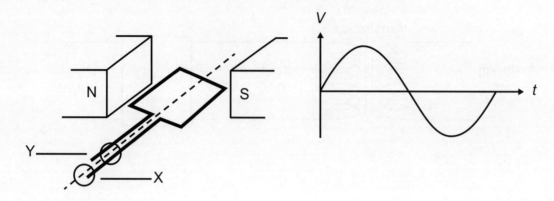

Which one of the following types of electrical output is most likely to be observed between the terminals labelled X and Y?

A.   constant magnitude DC voltage

B.   sinusoidal magnitude AC voltage

C.   fluctuating magnitude DC voltage

D.   constant magnitude AC voltage

Question 7

Which of the following correctly describes the purpose of an inverter in a solar power system?

A.   It increases the efficiency of PV cells.

B.   It converts sunlight into electrical energy.

C.   It stores excess energy produced by the PV cells.

D.   It converts DC electricity to AC electricity for grid connection.

SET 2

Question 1 (6 marks)

The diagram below shows a generator consisting of 200 loops of wire in a uniform magnetic field of 6.0 mT. The area within the loops is 0.03 m².

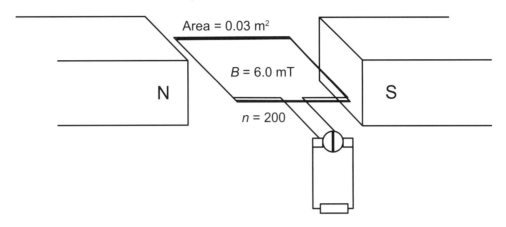

Area = 0.03 m²

B = 6.0 mT

N

S

n = 200

a. Calculate the maximum flux that can pass through the loops. Include a unit with your answer.　2 marks

b. The coil is rotated at a frequency of 20 Hz.

Calculate the magnitude of the average EMF produced by the generator.　2 marks

V

c. Another generator is rotated at 10 Hz and produces an average EMF of 3.5 V_{RMS} through a set of slip rings. Sketch the output of this generator on the axes below. Include a scale on the vertical axis.   2 marks

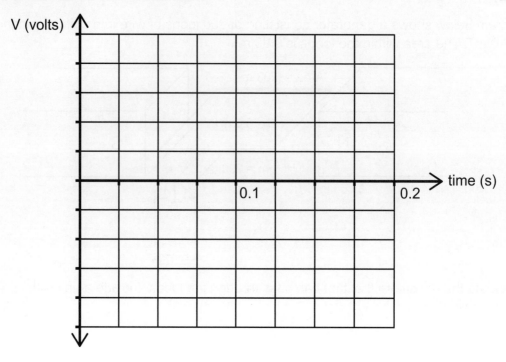

Question 2 (6 marks)

A small house is powered by a generator that produces 250 V_{RMS} through transmission lines that have a total resistance of 6.0 Ω, as shown in the diagram below. With just the lights and the TV switched on, the house draws 5.0 A and appliances operate normally.

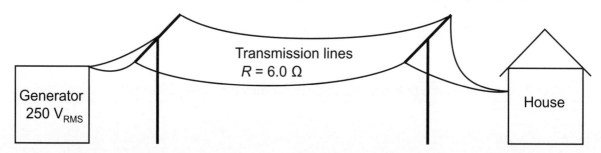

a. When the residents turn on their electric kettle and microwave, the house draws 10.0 A.

What voltage is provided to the house under these circumstances?   2 marks

	V

A 1:5 step-up transformer is installed at the generator and a 5:1 step-down transformer is installed at the house, as shown below.

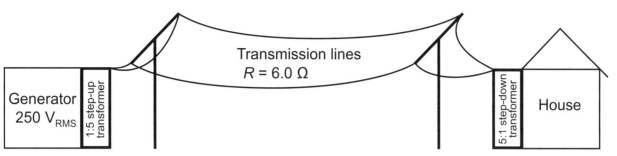

b. When the house draws 10.0 A, how much current flows in the transmission lines?   1 mark

A

c. 2500 W of power is produced by the generator.

What percentage of the power is lost in the transmission lines when the house draws 10.0 A?   3 marks

%

SET 3

Question 1 (6 marks)

A new generation system is being designed to supply electrical power to a country town several kilometres away from the generator. The system consists of a generator, two ideal transformers, T1 and T2, and a transmission line, as shown schematically below.

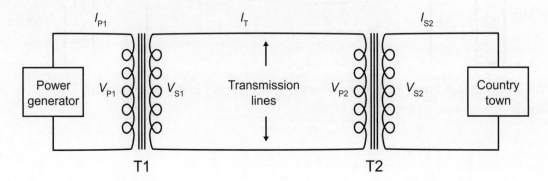

The generator is anticipated to supply 8.4×10^5 W of power at 240 V_{RMS} on the primary side of T1. The power is transferred via a step-up transformer, T1, using transmission lines with a total resistance of 3.0 Ω, and finally a step-down transformer, T2, with the same turns ratio as T1.

The town engineer is considering which turns-ratio to use for both transformers and is aware that the voltage to appliances in the town must not exceed 240 V_{RMS}.

a. What is the peak-to-peak voltage for an RMS voltage of 240 V?   1 mark

V

b. Show that I_{P1}, the anticipated current in the primary coil of transformer T1, is 3500 A_{RMS}.   2 marks

c.   The anticipated current in the primary coil of transformer T1 is 3500 A$_{RMS}$.
The town engineer is considering implementing a turns ratio of 100.

Calculate the power loss in the transmission line for the stated turns ratio.   3 marks

_____

_____

_____

_____

_____

	%

Question 2 (3 marks)

The diagram immediately below shows a schematic diagram of a conductor that is completely within a uniform magnetic field of 0.8 T. The length of the conductor is 12 cm. The magnitude of the current through the conductor is 0.6 A, in the direction shown on the diagram.

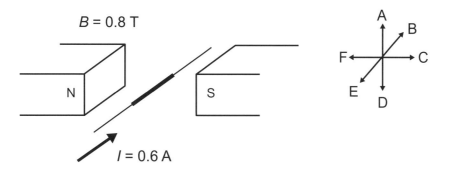

a.   The power supply is now disconnected from the conductor and a voltmeter is connected instead, as shown below. The conductor is now moved upwards at a constant velocity, $v = 0.15$ m s^{-1}.

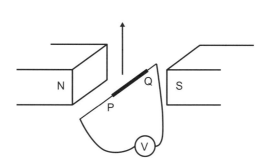

Calculate the EMF generated between P and Q, the two ends of the conductor.   2 marks

_____

_____

_____

_____

V

b. Determine the direction of the current in the conductor. Circle the correct answer in the list below.   1 mark

    from P to Q from Q to P There is no current.

Question 3 (5 marks)

The production line of a car parts factory uses a conveyor belt to transport parts from one section of the factory to another. The car parts are magnetic and could be detected using a detector system that comprises a conductor coil of wire and a voltage detector located overhead, as shown below. The conductor coil fits within the magnetic field of each car part.

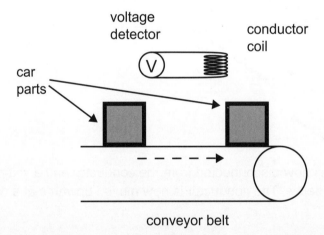

conveyor belt

The conductor coil comprises eight circular loops of copper wire with a diameter of 250 mm. The conveyor belt moves at a constant speed, and the time for the magnetic flux from the car parts to change from zero to maximum flux is $t = 0.65$ s. The conveyor belt speed can be varied from half to double the initial set speed.

a. The average EMF (ε) generated in the conductor coil must be at least 55 mV in order to detect a car part moving under the detector system.

Determine the minimum magnetic field strength at the location of the coil that would cause a car part to be detected by the detector system.   3 marks

	mT

A production engineer measures the magnetic field strength of the car part and finds that it is approximately 10% lower than the required field strength in **part a.**

b. Using the same equipment, suggest one way that the production engineer could make the car parts detectable. Explain why your suggestion would work.   2 marks

Question 4 (2 marks)

Why is it necessary to use an inverter in a rooftop solar installation?

SET 4

Question 1 (14 marks)

Lord Armstrong, a nineteenth-century British engineer, is believed to have been the first to use hydropower to light his home. He installed water turbines that drove electric generators to produce 4.5 kW of DC electric power at an output voltage of 55 V. This was the same voltage required for the electric arc lamps, located in his home about 120 m away, to operate normally. Copper cables were used to transmit the electric power directly, with a total transmission cable resistance of 0.3 Ω.

The power supply and transmission system may be modelled as a simple DC circuit, with a single electric arc lamp for analysis, as shown below.

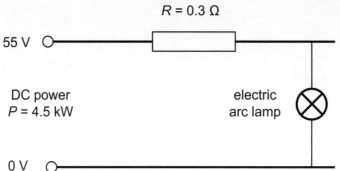

a. What is the size of the current in the transmission line?   2 marks

A

b. Calculate the total voltage drop across the transmission line.   2 marks

V

c. Determine the voltage available to the electric arc lamp. 2 marks

```
┌─────────────────────────┐
│                      V  │
└─────────────────────────┘
```

d. If the DC generator is replaced with an AC generator, and a suitable pair of step-up and step-down transformers are used, as shown below, the voltage drop across the transmission lines could be reduced. This would provide more voltage to the electric arc lamp. For simplicity, the turns ratio of the step-up transformer is the same as the turns ratio of the step-down transformer.

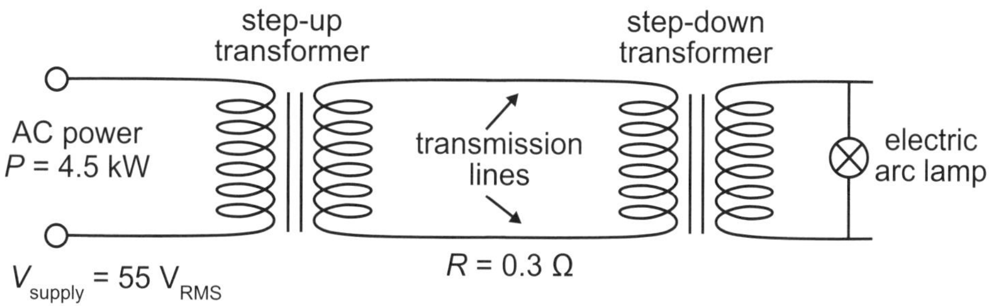

i. Using a turns ratio of 1:3 step-up, what is the expected value of the transmission line current? 1 mark

```
┌─────────────────────────┐
│                      A  │
└─────────────────────────┘
```

ii. Using a turns ratio of 1:3 step-up, what is the expected value of the voltage on the transmission line side of the step-up transformer? 1 mark

```
┌─────────────────────────┐
│                      V  │
└─────────────────────────┘
```

e. Determine the new voltage drop across the transmission line in this new set-up.   2 marks

_____

_____

_____

_____

	V

f. At the step-down transformer, what is the expected voltage across the primary side?   2 marks

_____

_____

_____

_____

	V

g. Using the same turns ratio, determine the expected voltage available to the electric arc lamp after the step-down transformer.   2 marks

_____

_____

_____

_____

	V

SET 5

Question 1 (11 marks)

The Asterisk Powerbank is a battery power storage system designed to provide backup power for small communities. The battery provides 25 kW of power at the output voltage of 230 V DC, and power lasts for 1 hour.

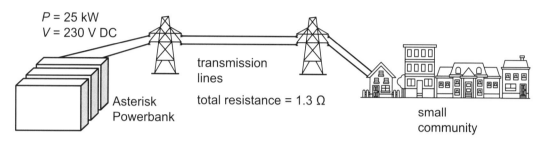

Figure 1

The planned installation for one small community involves transmitting the power via transmission lines that have a total resistance of 1.3 Ω.

a. Calculate the amount of power that will be lost during transmission. 3 marks

kW

b. It is decided to step up the voltage before transmission in order to reduce the power loss during transmission. However, stepping up the voltage of the Asterisk Powerbank cannot be achieved using a step-up transformer.

    i. If stepping up the voltage before transmission were possible, explain how this would reduce power loss. 2 marks

ii. Explain why this cannot be achieved with the DC output of the battery.   2 marks

_____

_____

_____

_____

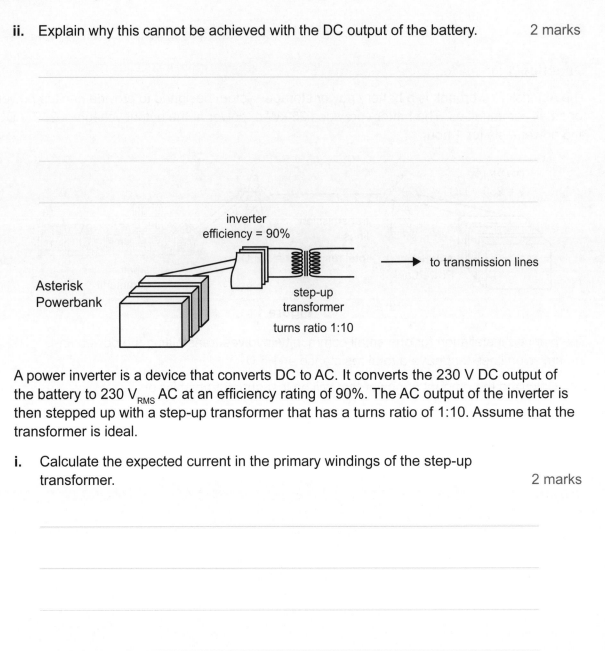

inverter
efficiency = 90%

Asterisk
Powerbank

to transmission lines

step-up
transformer

turns ratio 1:10

c. A power inverter is a device that converts DC to AC. It converts the 230 V DC output of the battery to 230 V_{RMS} AC at an efficiency rating of 90%. The AC output of the inverter is then stepped up with a step-up transformer that has a turns ratio of 1:10. Assume that the transformer is ideal.

i. Calculate the expected current in the primary windings of the step-up transformer.   2 marks

_____

_____

_____

A_{RMS}

ii. With the same transmission lines (total resistance = 1.3 Ω), what is the power loss in the transmission lines with the set-up shown in the diagram above?   2 marks

_____

_____

_____

W

Unit 4 | Area of Study 1   How has understanding about the physical world changed?

SET 1

Question 1

A spaceship, which has a length of 750 m when at rest, is travelling past a planet at $0.95c$ ($\gamma = 3.2$).

What is the length of the spaceship, as measured by a passenger on board?

A.   234 m

B.   750 m

C.   1500 m

D.   2400 m

Question 2

In a water tank, an oscillating motor creates ripples that are passed through a narrow gap, as shown in the diagram below.

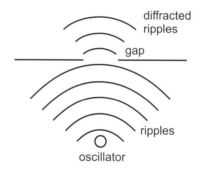

Which one of the following will result in an increase in the diffraction of the ripples?

A.   a narrower gap

B.   a shorter wavelength

C.   moving the oscillator closer to the gap

D.   increasing the frequency of the oscillator

Question 3

The diagram below depicts a water wave.

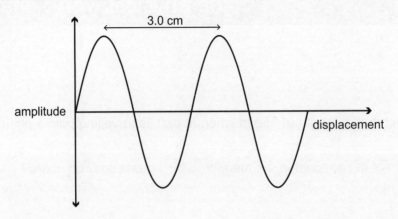

If the wave has a frequency of 300 Hz, which one of the following best gives the speed of the wave?

A.   900 m s^{-1}

B.   90 m s^{-1}

C.   9.0 m s^{-1}

D.   0.9 m s^{-1}

Question 4

A guitar string (fixed at both ends) is plucked and produces a fundamental frequency of 440 Hz.

Which one of the following resonant frequencies will also be present in the string?

A.   110 Hz

B.   220 Hz

C.   660 Hz

D.   880 Hz

SET 2

Question 1

Diffraction of waves may be observed when

A.   waves move from one medium to a different medium.

B.   there is a change in the speed of the wave.

C.   waves move through an aperture.

D.   waves collide against an obstacle at an angle.

Question 2

Deep-sea divers breathe a mixture of helium and air, known as trimix, so that they do not suffer from decompression sickness. A side-effect of this mixture is that their voice changes pitch due to a change in the speed of sound in trimix. A sound with frequency $f = 440$ Hz in air changes pitch to a frequency of $f = 1300$ Hz, although the wavelength remains constant in both mediums.

Given that the speed of sound in air is 340 m s^{-1}, the speed of sound in trimix is closest to

A.   115 m s^{-1}

B.   263 m s^{-1}

C.   1300 m s^{-1}

D.   1005 m s^{-1}

Question 3

Chester is given a monochromatic laser of an unknown wavelength. He carries out a double-slit interference experiment using the laser. Having determined that the distance between the two slits is $d = 250$ μm, he then projects the interference pattern onto a screen. The distance from the double-slit to the screen is 1.912 m, and the distance between the central maximum and the first maxima is $\Delta x = 3.1$ mm, as shown below.

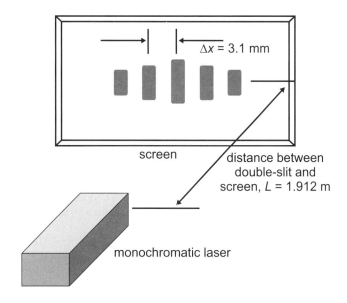

The wavelength of the laser is

A.   154 nm

B.   237 nm

C.   247 nm

D.   405 nm

Question 4

A beam of electrons passing through a thin layer of crystal produces a diffraction pattern, as shown below.

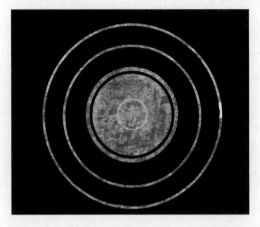

The kinetic energy of the individual electrons as they pass through the crystal is 36 eV. Calculate the de Broglie wavelength of the electrons.

A. 8.19×10^{-10} m

B. 5.76×10^{-10} m

C. 2.05×10^{-10} m

D. 1.28×10^{-10} m

SET 3

Question 1

The Australian Synchrotron produces high-intensity light using a beam of electrons travelling in a circular motion. The synchrotron accelerates the beam of electrons from rest to nearly the speed of light. In one particular run, the electrons pass a 1 m ruler that is placed parallel to the electrons' direction of travel.

From an electron's frame of reference, which one of the following 4 diagrams represents the graph of L, the length of the ruler, versus the speed of the electron?

A.

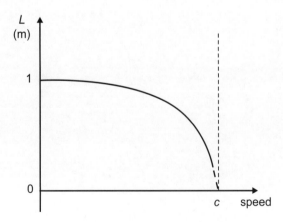

B.

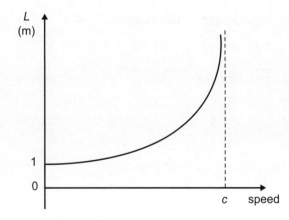

C.

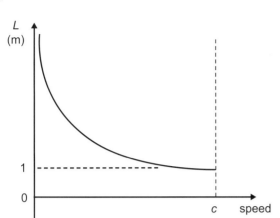

D.

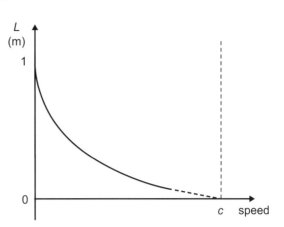

Use the following information to answer Questions 2 and 3.

The Beet FM radio station broadcasts at a frequency of 88.8 MHz. The Beet FM transmitter at the top of Mount Dandenong in Victoria has an output power of 12.5 kW.

Question 2

Which one of the following is closest to the wavelength of the radio signal from Beet FM?

A.   1.7 m

B.   2.5 m

C.   3.4 m

D.   6.8 m

Question 3

Assuming that the transmitter is ideal, the number of photons emitted per second is closest to

A. 3.4×10^{10}

B. 3.0×10^{18}

C. 2.1×10^{29}

D. 1.9×10^{37}

SET 4

Question 1 (7 marks)

a.   A speaker plays a tone of 500 Hz. Calculate the wavelength of the sound wave
produced. Take the speed of sound in air to be 340 m s^{-1}.   1 mark

b.   A second, identical speaker is placed beside the first and they both play the same tone, in
phase, with a wavelength of 0.760 m.

A student stands in front of the speakers at location L, as shown in the diagram below.
L is an equal distance from each speaker. The student then walks in a straight line to
location M.
M is 3.80 m from speaker 1 and 1.14 m from speaker 2.

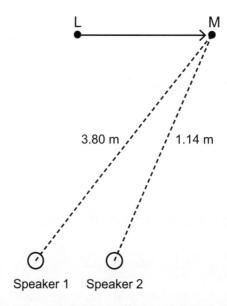

In the table below, describe how the student perceives the loudness of the sound
as they walk from location L to location M. Ignore echoes and reflected sound.   3 marks

	Perception of sound
At L	
From L to M	
At M	

c. In a room with the door open, a single speaker emits two tones together: one at 11 000 Hz and another at 1000 Hz. Two students, Ali and Bess, stand on the other side of the door, as shown below.

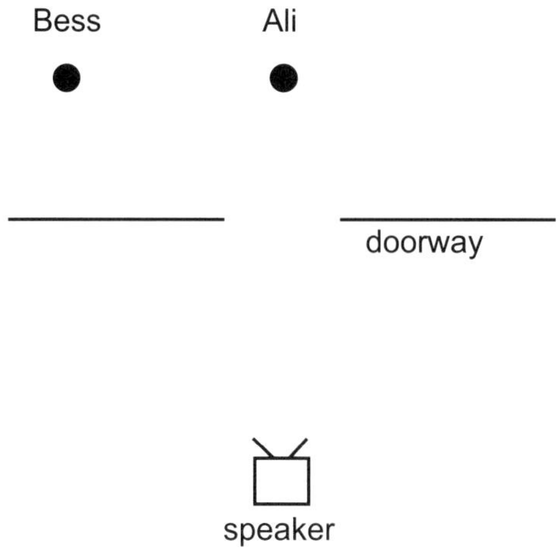

Bess

Ali

doorway

speaker

Ali hears both tones equally loudly.

Describe how Bess hears both tones. Use appropriate physics principles to support your answer.

3 marks

Question 2 (6 marks)

Shan is conducting an experiment to investigate the wave-like behaviour of light, using several monochromatic lasers of different wavelengths, some double-slit slides with varying slit separation (*w*) and a screen. She shines a laser beam through a double-slit slide and obtains a pattern on the screen. The pattern is a series of bright fringes, with no laser illumination between them, as illustrated below.

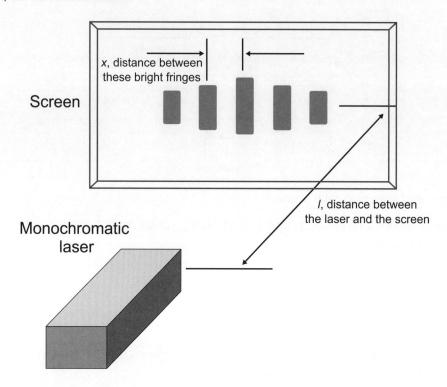

a. Using a double-slit slide with a slit separation of 10 μm, Shan investigates the effect of increasing the wavelength of the laser used.

State whether the distance between the fringes, *x*, will increase or decrease with increasing laser wavelength.   1 mark

b. Shan proceeds to investigate the effect of increasing the slit separation of the double-slit slide, using only a helium–neon laser with a wavelength of 633 nm.

State whether the distance between the fringes, *x*, will increase or decrease with increasing slit separation.   1 mark

c. Identify the name of the pattern on the screen. Circle the correct answer from the choices below.   1 mark

diffraction / interference

d. Explain how the pattern on the screen that is produced by the laser beam shining through a double slit supports the idea that light behaves as a wave.   3 marks

_____

_____

_____

_____

_____

_____

_____

Question 3 (10 marks)

In a particular thought experiment, Chris is on board a moving train that is travelling at $0.98c$ relative to Deb, as shown below.

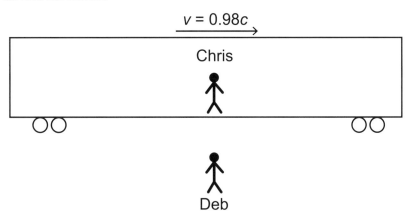

a. Chris measures the train to be 10 m long.

How long would Deb measure the train to be?   2 marks

_____

_____

_____

_____

_____

_____

Some time later, the train is travelling at a constant velocity of 2.97×10^8 m s^{-1}. Chris sees Deb drop a ball. From Chris' perspective, it takes 4.0 s to reach the ground.

b. How long does Deb see it take for the ball to reach the ground? 2 marks

_____

_____

s

c. Both Chris and Deb are convinced that they are at absolute rest because no net force acts upon them.

Chris claims that he is at rest relative to the universe because he is in an inertial frame of reference.

Deb claims that she is at rest relative to the universe because she is not moving.

Who is correct? Choose one option below and explain your answer by referring to Einstein's first postulate. 3 marks

Chris / Deb / both / neither

_____

_____

_____

_____

d. When four hydrogen nuclei are combined into one helium nucleus in the Sun, 25 MeV of energy is released.

Calculate how much mass is lost when this happens and identify how many protons or neutrons (if any) are destroyed in this process. 3 marks

_____

_____

_____

_____

SET 5

Question 1

Gini is a mission controller on a space station. She observes Jordan in a spaceship travelling past her at a relative speed of $0.866c$ ($\gamma = 2.00$). Both Gini and Jordan each hold up a metre ruler parallel to the direction of their travel past each other. Due to relativistic effects, Gini measures Jordan's ruler as 0.5 m, a phenomenon known as length contraction.

What will be the measurement that Jordan makes of Gini's ruler?

A.   0.5 m

B.   1.0 m

C.   1.5 m

D.   2.0 m

Question 2

During experiments into the photoelectric effect, a stopping voltage is commonly applied to the anode.

This is done to

A.   amplify the photocurrent.

B.   give an indication of the maximum kinetic energy of the electrons.

C.   give an indication of the work function of the metal.

D.   give an indication of the intensity of the light.

Question 3

Light may be modelled as a stream of particles known as photons.

Which one of the following experiments demonstrates the particle-like nature of light?

A.   light emitted during electron transitions in an emission spectrum experiment

B.   incident light on a photocell experiment, demonstrating the photoelectric effect

C.   light absorbed by valence electrons during an absorption spectrum experiment

D.   all of the above

Question 4

In 1902, Philipp Lenard's experiments with photoelectrons, in which he shone light onto a metal target, provided the key evidence for the particle-like nature of light. Which of the following observations about the experiment supports the particle model of light?

A. A higher intensity light source results in photoelectrons with higher kinetic energy, and thus a higher stopping voltage.

B. A light source at any frequency still produces photoelectrons, but lower frequencies may experience a time delay.

C. A higher frequency light source results in photoelectrons with higher kinetic energy, and thus a higher stopping voltage.

D. A lower intensity light source results in photoelectrons with lower kinetic energy, and thus a lower stopping voltage.

SET 6

Question 1

X-rays directed through very thin layers of crystal form a diffraction pattern with the same fringe spacing as the diffraction pattern formed by an electron beam passing through the same layers of crystal.

Which conclusion can be drawn from this observation?

A. The X-rays interact with the thin layers of crystal in the same particle-like manner as the electrons, hence forming the same pattern as the electron beam.

B. The X-rays and the electron beam behave in a wave-like manner and have the same frequency, hence the same pattern.

C. The electron beam and the X-rays behave in a wave-like manner and have the same wavelength, hence the same pattern.

D. The electron beam travels through the thin layers of crystal in a particle-like manner, hence forming the same pattern as the X-rays.

Question 2

An emission spectrum experiment detected a photon of light with a wavelength of 560 nm.

The energy of the photon is closest to

A. 3.55 eV

B. 3.55 MeV

C. 2.22 eV

D. 2.22 MeV

Question 3

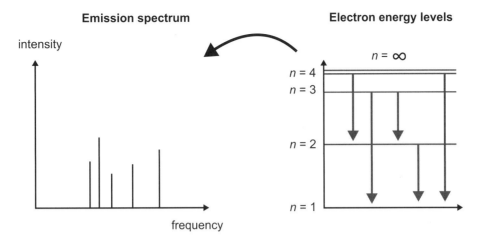

The emission spectrum of elements (above left) shows discrete lines corresponding to the different light frequencies emitted by electrons as they transition between energy levels (above right).

Which one of the following statements explains this behaviour?

A. Electrons orbit as matter waves, with energy levels corresponding to the intensity of the light emitted.

B. Electrons orbit as particles in circular paths, with a radius proportional to the temperature of the material.

C. Electrons orbit in wave-like orbital paths that increase in frequency at higher energy levels.

D. Electrons orbit as matter waves, with energy levels corresponding to harmonics (integer multiples) of standing waves.

Question 4

A cathode ray tube is used to accelerate electrons to a constant speed. These electrons pass through a thin slice of a salt crystal and form a diffraction pattern on a phosphor screen on the wall of the cathode ray tube.

Which one of the following best relates the behaviour of the electrons and the observations made in the cathode ray tube?

	Electrons diffracting through the salt crystal	Electrons arriving at the phosphor screen
A.	electrons behave in a wave-like manner	electrons behave in a wave-like manner
B.	electrons behave in a particle-like manner	electrons behave in a particle-like manner
C.	electrons behave in a wave-like manner	electrons behave in a particle-like manner
D.	electrons behave in a particle-like manner	electrons behave in a wave-like manner

SET 7

Question 1 (3 marks)

Below is the partial energy level diagram for mercury.

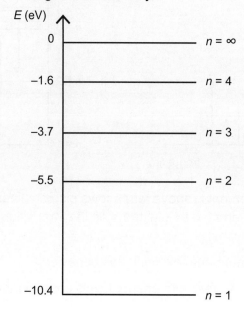

Calculate the energy of a photon with a frequency of 9.42×10^{14} Hz and draw a single arrow on the diagram to show how this photon could be emitted from a mercury atom.

3 marks

 eV

Question 2 (8 marks)

a. Electrons are accelerated from rest by an electron gun at 15.0 V. Show that the de Broglie wavelength of these electrons is 3.2×10^{-10} m. Relativistic effects can be ignored. 3 marks

b. What frequency of electromagnetic radiation would be required to produce a diffraction pattern that is identical to that produced by the electrons described in **part a.**? 2 marks

Hz

c. A flea (200 µg) can jump at 2.0 m s^{-1}. How many violet photons ($\lambda = 400$ nm) are required to produce the same momentum as that of a jumping flea? 3 marks

SET 8

Question 1 (6 marks)

In a photoelectric effect experiment, ultraviolet light at a frequency of 8.0×10^{14} Hz is incident on a sample of lithium metal (work function = 2.9 eV).

A stopping voltage of 0.45 V is recorded.

a.   Calculate the threshold frequency for this metal.   2 marks

 Hz

Another metal is tested and the results are shown on the following graph.

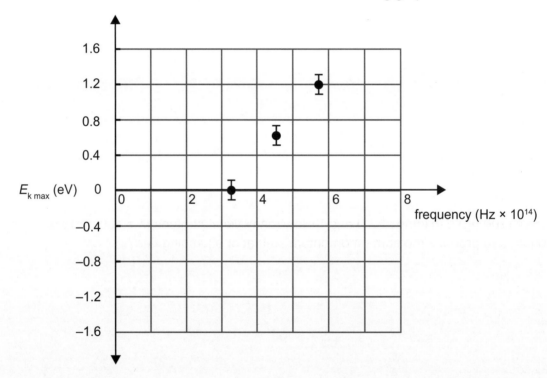

b.   Use the graph on the previous page to calculate the value of Planck's constant.
Clearly show on the graph how you reached your answer.   3 marks

	eV s

c.   Use the graph to calculate the stopping voltage that would be recorded
when light of 5.0×10^{14} Hz is incident on this metal. Clearly show on the graph
how you reached your answer.   1 mark

	V

Question 2 (7 marks)

The energy level diagram for an element, Z, is shown below.

$n = 4$ ———————————————— 2.18 eV

$n = 3$ ————●———————— x eV

A

$n = 2$ ————↓———————— 0.47 eV

$n = 1$ ———————————————— 0 eV

a.   The arrow labelled A represents one of the possible energy level transitions that element Z
can undergo when excited to the $n = 3$ state.

On the diagram, draw the other **two** possible transitions.   2 marks

b. The energy level transition shown by the arrow labelled A is associated with the emission of a photon of momentum 7.1×10^{-28} N s.

Determine the value for x of the $n = 3$ energy level of element Z.   3 marks

_____

_____

_____

_____

eV

c. Explain why a beam of photons with energy 0.55 eV, incident on a gaseous sample of element Z, will not be absorbed.   2 marks

_____

_____

_____

_____

SET 9

Question 1 (3 marks)

The photoelectric effect involves shining particular frequencies of light at varying intensities onto a metal plate and measuring the rate and the kinetic energy of the ejected electrons. Experiments such as these were used by Einstein and others to support the particle model of light.

Describe the results of photoelectric effect experiments that would be expected under the **wave model** of light.

Question 2 (6 marks)

A teacher is demonstrating the photoelectric effect in class by bombarding a metal sample with monochromatic light in an evacuated tube. She records the following single piece of data: when the frequency of the incident light is 6.45×10^{14} Hz, she finds the stopping voltage to be 1.47 V.

a. Determine the energy of the most energetic photoelectrons emitted from the metal sample. Give your answer in joules. 1 mark

	J

b. Use your result to determine the work function of the metal sample. Give your
answer in eV.   3 marks

_____

_____

_____

_____

_____

eV

c. The teacher now changes the light source to one with a longer wavelength, and
photoelectrons are still emitted.

Will the stopping voltage increase, decrease or stay the same?   1 mark

_____

d. The teacher now increases the intensity of the light.

Will the stopping voltage increase, decrease or stay the same?   1 mark

_____

Question 3 (5 marks)

A research scientist is using diffraction techniques to investigate the structure of a crystal. The
diagram below shows a crystal being bombarded and producing a diffraction pattern.

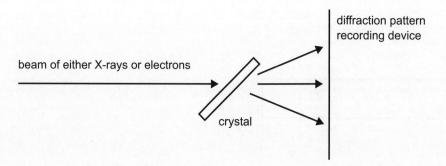

In one experiment the scientist chooses to use a beam of X-rays, and in a second experiment
he chooses to use a beam of electrons to strike the crystal.

In each case the X-rays and electrons have the same wavelength of 2.0×10^{-10} m.

a. Determine the energy of a single X-ray photon used in the experiment.   2 marks

J

b. Determine the kinetic energy of a single electron used in the experiment.   3 marks

J

SET 10

Question 1 (3 marks)

Jon and Muriel are investigating the photoelectric effect.

Light is passed through a yellow filter and is incident on a metal plate. By adjusting a variable DC voltage, Jon and Muriel gather enough data to be able to plot the following graph of the photocurrent against voltage. The voltage at which the photocurrent goes to zero is $V = V_{stop}$ and the photocurrent when the variable voltage is zero is $I = I_{yellow}$.

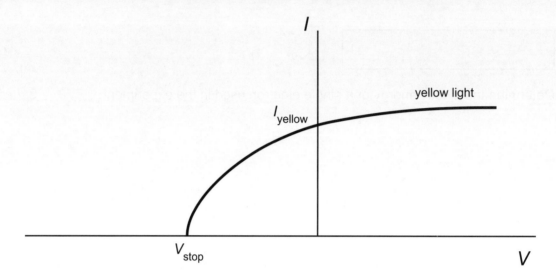

The intensity of the light source is increased so that the light power output of the yellow light incident on the metal plate increases.

a. On the axes above, sketch the graph of the photocurrent expected, using a solid line.   1 mark

b. The filter is changed to a green filter. The intensity of the light source is adjusted so that the light power output of the green light incident on the metal plate is the same as that of the yellow light at the beginning of the experiment, which produced the initial graph on the axes.

On the same axes, sketch the graph of the photocurrent expected, using a dashed line.   2 marks

Question 2 (3 marks)

The electron energy level diagram for the element lithium is shown below. As the electrons transition from a higher energy level to a lower one, they emit photons of particular wavelengths, which show up as lines on the emission spectrum.

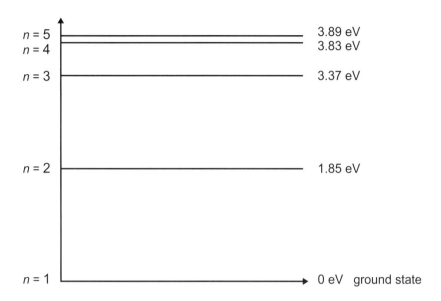

a. One of the emission lines on the spectrum for lithium has a wavelength of 627 nm.

Show that the energy of the photons that produce this emission line is 1.98 eV.   2 marks

b. On the diagram above, draw an arrow that represents the transition that will emit the photon with a wavelength of 627 nm. Clearly show where the arrow starts and ends.   1 mark

SET 11

Question 1 (7 marks)

Moh and Div are experimenting with different coloured filters as they investigate the photoelectric effect. Their apparatus is shown below.

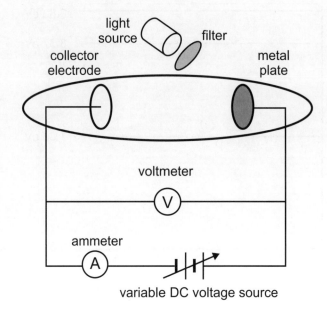

The aim of the experiment is to determine the threshold frequency of the metal plate and the work function of the metal. Moh and Div obtain the following results for the maximum kinetic energy ($E_{k\,max}$) of the emitted photoelectrons versus the frequency of the light incident on the metal plate.

The results of their investigation are shown in Table 1 below.

Table 1

f ($\times 10^{14}$ Hz)	$E_{k\,max}$ (eV)
3.90	0.00
4.15	0.00
4.89	0.05
5.15	0.19
5.75	0.42
6.25	0.62

a. On the axes provided below:

- Plot the data presented in Table 1.

- Include appropriate scales on each axis.

- Draw a line of best fit that is suitable for determining the threshold frequency of the metal plate.   4 marks

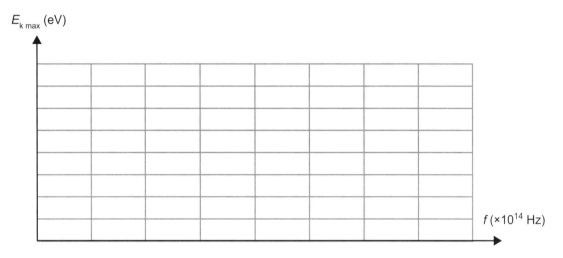

$E_{k\,max}$ (eV)

$f\,(\times 10^{14}$ Hz$)$

b. State the threshold frequency that can be derived from the results of the experiment.   1 mark

Hz

Moh and Div calculate the gradient of their graph to be 4.1×10^{-15} eV s.

c. Using this value and the information on your graph, determine the work function, W, of the metal used in this photoelectric experiment.   2 marks

eV

Question 2 (3 marks)

A teacher is demonstrating how diffraction patterns from X-rays and electrons that are passed through thin metal foils might appear similar to each other even though the means of producing them are different. The pattern from an electron experiment is combined with one from an X-ray experiment, shown below. Note that both patterns shown are to the same scale.

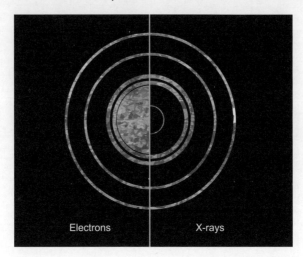

Electrons   X-rays

The X-rays that produce the pattern have a momentum of 2.47×10^{-23} N s.

Calculate the kinetic energy of the electrons in this experiment.

_____

_____

_____

_____

	J

Question 3 (3 marks)

The emission spectrum of an element shows discrete lines corresponding to the different light frequencies emitted by electrons as they transition between energy levels. In one experiment involving a particular element, electrons transitioned from level $n = 3$ back to ground state ($n = 1$).

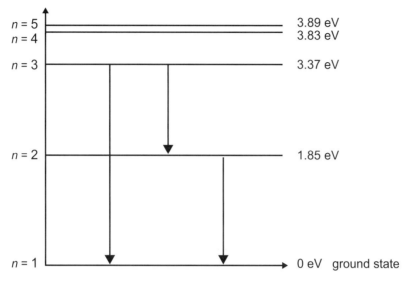

The observed emission spectrum that was produced displayed spectral lines at the frequencies indicated below.

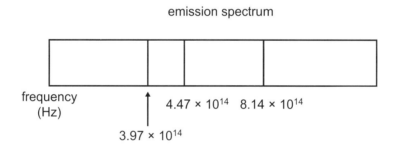

Elise, a physics student, notes that the first reported spectral line, with frequency $f = 3.97 \times 10^{14}$ Hz, must be a mistake.

Explain why this frequency could not be observed when the electrons of this element transitioned from level $n = 3$ back to ground state and support your answer with calculations.

Unit 4 | Area of Study 2   How is scientific inquiry used to investigate fields, motion or light?

SET 1

> *Use the following information to answer Questions 1 and 2.*
>
> Some students are investigating how temperature affects the spring constant, k, of various brands of elastic.

Question 1

Which one of the following would be a discrete independent variable?

A.   the temperature of the elastic

B.   the brand of the elastic

C.   the extension of the elastic

D.   the spring constant, k, of the elastic

Question 2

Which one of the following might the students choose to be a controlled variable?

A.   the brand of elastic

B.   the initial length of the elastic

C.   the temperature of the elastic

D.   the spring constant, k, of the elastic

Question 3

Some students are investigating the behaviour of a pendulum.

Which one of the following could potentially be a dependent variable?

A.   the mass of the weight at the end of the string

B.   the length of the string

C.   the angle of release

D.   the period of oscillation

Question 4

Ally and Billy carry out an experiment to determine the relationship between the period of a pendulum and the length of the pendulum. They plot their data, as shown in the graph below.

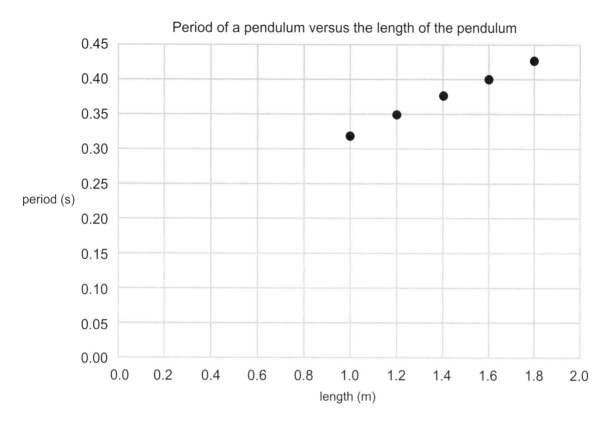

Ally says to Billy, 'By linearly extrapolating the graph, we can see that if we reduce the length of the pendulum to 0.5 m, the period of the pendulum will reduce to 0.25 s.' Billy disagrees with Ally's prediction, saying, 'That is an improper use of the graph.'

Which one of the following is the **most correct** reason that Billy could provide?

A. It is never correct to extrapolate a graph because it is an unscientific practice.

B. There is no data point at the length of 0.5 m and actual measurements have to be made.

C. The relationship cannot be assumed to continue to be linear within the region of 0.5 m.

D. The textbook clearly says the relationship is not linear, therefore you should not extrapolate.

Question 5

Which one of the following errors cannot be reduced by having an experiment repeated by different experimenters using the same set of apparatus and equipment, and taking the average value of all measurements?

A. random errors

B. systematic errors

C. outliers

D. personal errors

SET 2

Question 1

Shawn takes several measurements of the natural frequency of a violin string using a frequency analyser and obtains the following readings on the instrument.

442.4 Hz, 441.1 Hz and 440.9 Hz

Which one of the following is the **best representation** of the average and the uncertainty of these readings?

A.   441.47 ± 0.93 Hz

B.   441.50 ± 0.90 Hz

C.   441.5 ± 0.9 Hz

D.   441 ± 1 Hz

Question 2

Which one of the following statements does **not** correctly describe a hypothesis within the framework of the scientific method?

A.   A hypothesis is a tentative explanation for an observed phenomenon.

B.   A hypothesis may be supported by experimental evidence.

C.   A hypothesis is useful only if it is supported by experimental evidence.

D.   A hypothesis may be disproved by experimental evidence.

Question 3

An experiment on a frictionless air track, represented in the diagram below, is conducted to determine the relationship between a net force provided by a constant falling mass, the mass of a cart travelling on the air track and the acceleration of the cart.

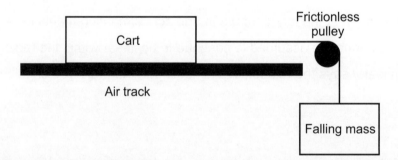

Which one of the following statements correctly classifies the variables involved?

A.   Acceleration is the dependent variable, the net force is the independent variable and the mass of the cart is the controlled variable.

B.   Acceleration is the independent variable, the net force is the dependent variable and the mass of the cart is the controlled variable.

C.   Acceleration is the dependent variable, the net force is the controlled variable and the mass of the cart is the independent variable.

D.   Acceleration is the independent variable, the net force is the controlled variable and the mass of the cart is the dependent variable.

Question 4

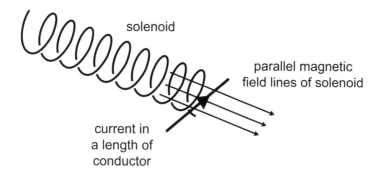

Angela is carrying out an experiment to determine the magnetic field strength of a solenoid, using $F = BIl$, by measuring the force on a length of conductor that is perpendicular to the magnetic field at one end of the solenoid, as shown above. The current in the conductor is modified using a variable resistor.

Which one of the following options correctly describes the variables in the experiment?

A. The current is the independent variable, the length of the conductor is the controlled variable and the force on the conductor is the dependent variable.

B. The magnetic field strength is the dependent variable, the current is the independent variable and the force on the conductor is the derived quantity.

C. The magnetic field strength is the independent variable, the force on the conductor is the dependent variable and the current is the controlled variable.

D. The current is the independent variable, the force on the conductor is the controlled variable and the length of the conductor is the dependent variable.

Question 5

Alex and Kate are experimenting with a frictionless air track. A cart (mass M = 480 g) is attached to a falling mass (mass m = 80 g) by a massless cable. The falling mass is allowed to free fall under the influence of gravity. The cable runs over a frictionless pulley and air resistance is negligible.

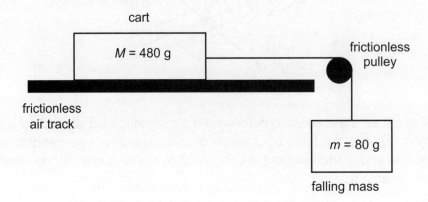

Alex and Kate measure the acceleration of the cart using an onboard accelerometer, the mass of which is included in the mass of the cart.

After two runs with the apparatus, the frictionless pulley becomes twisted. It now exerts a constant friction force of 0.15 N on the motion of the cable.

Which one of the following best explains how this constant friction force will affect the measurement of the acceleration of the cart, in terms of accuracy and precision?

A.   The constant friction force will affect the accuracy only.

B.   The constant friction force will affect the precision only.

C.   The constant friction force will affect both the accuracy and the precision.

D.   The constant friction force will affect neither the accuracy nor the precision.

SET 3

Question 1 (8 marks)

Some students are investigating a step-down transformer, attempting to find the ratio of turns in the primary coil to the secondary coil.

The students apply a voltage to the primary coil, which is measured using a digital multimeter (± 0.005 V). Uncertainty in the primary coil measurements may be ignored. The students use an analogue voltmeter marked in 1 V increments to record the voltage across a load resistor connected to the secondary coil. Their data is shown in the table below.

Primary coil (volts)	Secondary coil (volts)
0.10	1.8
0.20	4.3
0.30	5.9
0.40	7.7
0.50	10.1

a. Record the data from the table onto the grid below. Include realistic uncertainty (error) bars, labels, scales and units. Draw a line of best fit.   5 marks

b. Use the error bars on the graph to calculate the highest and lowest possible values for the ratio $\dfrac{\text{secondary coil voltage}}{\text{primary coil voltage}}$.   3 marks

Highest possible value:

Lowest possible value:

Question 2 (11 marks)

As part of a practical investigation, a group of students decides that they will release a ball bearing into a tall cylinder of oil and record several values of the variable t, the time it takes for the ball bearing to hit the bottom of the cylinder. They will do this for different depths of oil in the cylinder, the variable x.

Their initial hypothesis is that the greater the depth of oil, the longer the time it will take for the ball bearing to reach the bottom of the cylinder.

a.   For this experiment, state the dependent and independent variables.   2 marks

The dependent variable is: _____

The independent variable is: _____

b.   Give an example of a quantity that should be controlled and kept constant, as far as possible, as the students collect data.   1 mark

_____

c.   Explain why repeat measurements of a single trial within the experiment should be made.   1 mark

_____

_____

The data recorded by the group is shown in the table below. The data is the average of three measurements (which are not shown).

x, depth of oil (m)	t, time to hit bottom (s)
0.10	0.92
0.20	1.35
0.80	2.75
1.20	3.35
2.00	4.35
3.00	5.20

A graph of the data is shown below, along with the uncertainties associated with the measurements.

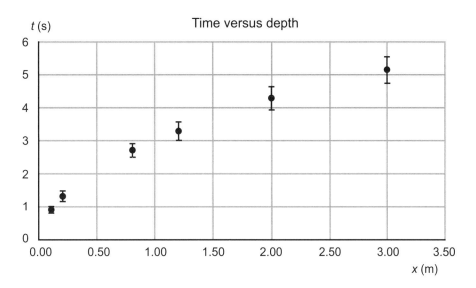

d. From the graph, estimate the absolute error in the timing measurement when the depth of oil is 3.00 m.

1 mark

	s

e. A student in the group suggests that the data could be better modelled by a non-linear relationship.

Of the options below, which would be the best proportionality to model the data? Circle your answer.

1 mark

-   Time taken to fall is inversely proportional to the depth of oil:
    $t \propto \dfrac{1}{x} \Rightarrow t = \dfrac{k}{x}$.

-   Time taken to fall is proportional to the square of the depth of oil:
    $t \propto x^2 \Rightarrow t = kx^2$.

-   Time taken to fall is proportional to the square root of the depth of oil:
    $t \propto \sqrt{x} \Rightarrow t = k\sqrt{x}$.

f. In order to test the choice of model to fit the data, a new variable, z, is to be computed from each value of x, using the model of your choice from **part e.**

Complete the second column in the table on the next page according to the model you selected:

2 marks

-   $z = \dfrac{1}{x}$
-   $z = x^2$
-   $z = \sqrt{x}$

x, depth of oil (m)	z, new variable to test the model chosen	t, time to hit bottom (s)
0.10		0.92
0.20		1.35
0.80		2.75
1.20		3.35
2.00		4.35
3.00		5.20

g. To test the fit of the model chosen, the data in the table above is to be plotted on a graph.

On the grid provided below, plot the data from the table. The *y*-axis scale has been added for you. You will need to choose the scale for the *z* variable.   2 marks

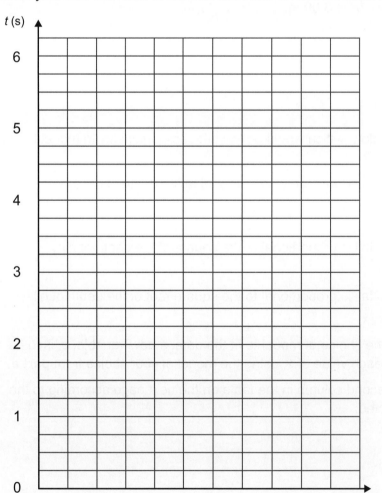

h. Does your graph support or not support your choice in **part e.**? Explain why.   1 mark

_____

_____

SET 4

Question 1 (10 marks)

The Kundt's tube is an apparatus that uses the standing wave phenomenon to determine the speed of sound in a gas. As shown in the diagram below, it comprises a clear plastic tube with a loudspeaker at one end and a sliding wall at the other end that can vary the length, L, of the section in between. Talcum powder is scattered along the length of the tube.

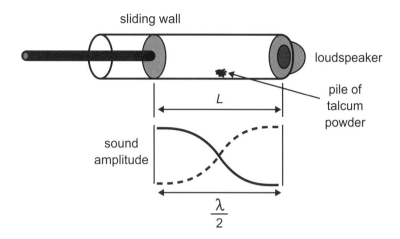

When the loudspeaker plays a pure tone sound, the sliding wall is moved until the sound volume reaches maximum loudness, indicating that L matches a resonant frequency. The talcum powder provides visual confirmation that the resonant frequency is fundamental because it forms a pile in the middle of the enclosed section, indicating that there is a node in the middle.

L corresponds to half the wavelength of the fundamental frequency, $\frac{\lambda}{2}$.

A class of students collect the following data for a few pure tone frequencies, f, and the length, L.

Frequency, f (Hz)	Length, L (cm)	$\dfrac{1}{L}$
165	109.1	0.92
330	52.9	1.89
587	29.2	3.42
880	19.3	5.18
1175	14.4	6.95

a. On the grid provided below: 5 marks

- Include scales and units on each axis.
- Plot the data from the table.
- Draw a line of best fit.

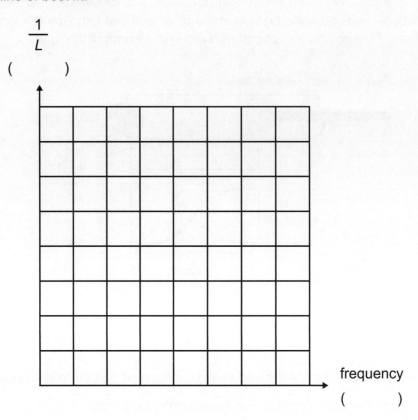

$\frac{1}{L}$

(   )

frequency

(   )

b. Determine the gradient of the line of best fit drawn in **part a.** Show your working. 2 marks

_____

_____

_____

_____

| m s⁻¹ |

c. Use the value of the gradient calculated in **part b.** to calculate the speed of sound in air. Show your full working.   3 marks

	m s^{-1}

Question 2 (4 marks)

In preparation for a physics experiment, Adrienne uses a mass balance to weigh five 50g slotted masses to ensure that they are suitable for the experiment. She finds that the average mass is 54.88 g. Later, she finds that the balance is poorly calibrated, resulting in all objects measuring 4.52 g more.

a. Identify the type of error that is caused by the poor calibration of the balance.   1 mark

b. **i.** Circle the aspect of Adrienne's measurements of the masses that will be affected by this type of error.   1 mark

accuracy only   precision only   both accuracy and precision

ii. Explain your choice in **part b.i.**   2 marks

● Worked solutions

Unit 3 | Area of Study 1   How do physicists explain motion in two dimensions?

SET 1

Question 1

Answer: A

Explanatory notes

Kinetic energy is gained as gravitational potential energy is lost:

$$E_k + E_g = E_{total}$$
$$E_k = -mg\Delta h + E_{total}$$

Hence

$$\Delta E_k = \Delta E_g$$
$$\Delta E_k = mg\Delta h$$

which is a linear relationship between E_k and Δh, with a maximum value of E_k when $h = 0$.

Question 2

Answer: B

Explanatory notes

At the bottom of its swing:

$$T - mg = \frac{mv^2}{r}$$
$$T = \frac{mv^2}{r} + mg$$
$$= \frac{0.06 \times 2.3^2}{0.4} + 0.06 \times 9.8$$
$$= 1.38 \text{ N}$$

Question 3

Answer: C

Explanatory notes

Rearranging the formula for centripetal force, $F_c = \frac{mv^2}{r}$, the speed of the airliner is given by $v = \sqrt{\frac{F_c r}{m}} = \sqrt{\frac{125\,500 \times 690}{15000}} = 76.0 \text{ m s}^{-1}$.

SET 2

Question 1

Answer: D

Explanatory notes

According to Newton's third law of motion, the force that Erica exerts on the seat is equal in magnitude but opposite in direction to the normal force exerted by the seat on Erica. The normal force, which is directed upwards, has a magnitude given by $N = m(g - a)$, where a is the downward acceleration. Substituting the values gives $N = 55(9.8 - (-3)) = 704$ N. Hence, the force that Erica exerts on the seat is 704 N, acting downwards.

TIPS

» For questions involving the normal force of a lift (or a lift-like vehicle, such as the ride in this question), it is worth remembering that the normal force is the weight experienced by the occupant. Thus, when the lift is accelerating downwards, the normal force is reduced and the occupant feels 'lighter'; whereas when the lift is accelerating upwards, the normal force is increased and the occupant feels 'heavier'.

» When writing a vector equation for the net force of an object, always use the direction of motion as positive.

Question 2

Answer: B

Explanatory notes

Erica has no kinetic energy at the beginning of her ride, nor at the end of the ride. Hence, the difference in her total mechanical energy is entirely the difference in her gravitational potential energy between the ground level and the highest level. Thus, the net work done on her is the same as increasing her potential energy; that is, $W = \Delta E_g = mg\Delta h = 55 \times 9.8 \times 20 = 10\,780$ J.

It should be noted that when Erica accelerated upwards, and while she was in motion, additional work was being done on her by the ride, but this is cancelled out as she slows down to a stop.

Question 3

Answer: C

Explanatory notes

On a distance–time graph, acceleration is represented by a parabola, and zero speed is represented by a horizontal line.

Question 4

*Answer: **A***

Explanatory notes

The three forces may be arranged in a vector diagram as a right-angled triangle, because forces that result in zero net force form a closed shape.

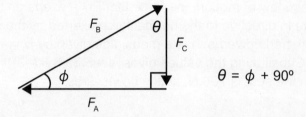

$$\theta = \phi + 90°$$

The angle θ is the sum of $90°$ and ϕ, the angle between F_A and F_B.

$$\phi = \cos^{-1}\frac{F_A}{F_B} = \cos^{-1}\frac{12}{13} = 22.62°$$

$$\theta = 90° + 22.62° = 112.6°$$

SET 3

Question 1

*Answer: **C***

Explanatory notes

The tension of the rope is providing the centripetal force that enables the ball to execute uniform circular motion. This tension force cannot exceed the breaking strength of the rope; hence, $F_c \leq 750$ N.

Since $F_c = \dfrac{mv^2}{r}$, then $v = \sqrt{\dfrac{F_c r}{m}} = \sqrt{\dfrac{750 \times 2.3}{12}} = 12.0$ m s^{-1}.

Question 2

*Answer: **B***

Explanatory notes

Decreasing the radius while maintaining the same speed requires a higher centripetal force to maintain the circular motion. The higher centripetal force will exceed the breaking strength of the rope, causing it to break. Each of the other options results in the ball requiring a lower centripetal force.

Question 3

Answer: **C**

Explanatory notes

Option A is incorrect because the total energy of the mass is maximum at zero extension, but reduces to zero at maximum extension because gravitational potential energy is zero and kinetic energy is also zero.

Options B and D are incorrect because the kinetic energy of the mass (which is also the same as the kinetic energy of the system) is maximum halfway between zero and maximum extension.

This means that option C is the correct option.

 TIP

» Be careful to distinguish between the 'total energy of the system' and the 'total energy of the mass', as the system energy is the sum of the mass energy and the spring energy; that is, $E_{system} = E_{mass} + E_{spring}$.

Question 4

Answer: **D**

Explanatory notes

Options A and B are incorrect, as the aeroplane is accelerating centripetally due to the centripetal force provided by the horizontal component of the lift of the aeroplane's wings. Option C is incorrect because the thrust of the aeroplane's engine is directed tangentially to the circle of its flight path. Option D is the only correct description of the aeroplane's motion.

SET 4

Question 1

Answer: **B**

Explanatory notes

The cart and falling mass may be considered to be a single body with a total mass of $m_{total} = 0.56$ kg (480 g + 80 g = 560 g). The net force on this body is the weight force on the falling mass, $F_{net} = F_w = mg = 0.08 \times 9.8 = 0.784$ N. Thus, the acceleration is

$$a = \frac{F_{net}}{m_{total}} = \frac{0.784}{0.56} = 1.4 \text{ m s}^{-2}.$$

Question 2

Answer: D

Explanatory notes

Option A is incorrect because inertia is not a force; inertia, or a body's resistance to change of its uniform motion, is equivalent to mass. Options B and C are incorrect because there is no force acting on the passengers towards the rear of the bus. Option D is correct because the floor of the bus exerts a small friction force on the feet of the passengers, but their bodies remain still as a result of inertia; hence, they fall backwards relative to the bus.

SET 5

Question 1a.

Worked solution

According to Newton's third law of motion, the force exerted on box A by box B is equal in magnitude but opposite in direction to the force exerted on box B by box A. Since the force on box B by box A is due to the weight of box A, the magnitude of the force on box A by box B is $m_A g = 3 \times 9.8 = 29.4$ N.

Since the direction of the force on box B by box A is *down*, the direction of the force on box A by box B is *up*.

Mark allocation: 2 marks

- 1 mark for the correct magnitude of the force
- 1 mark for the correct direction of the force

Question 1b.

Worked solution

Taking down as positive, and considering the forces on box A: the net force on box A is the sum of its force due to gravity and the force on box A by box B. Thus $\Sigma F = ma = 3 \times 3.5 = 10.5$, and $\Sigma F = W - F_{\text{on A by B}} = 10.5 \Rightarrow F_{\text{on A by B}} = W - 10.5 = 29.4 - 10.5 = 18.9$ N, and the direction of the force is *down*. Applying Newton's third law, the magnitude of the force on box B by box A is 18.9 N acting downwards.

Mark allocation: 3 marks

- 1 mark for equating the net force on box A to the sum of its force due to gravity and the normal force exerted by box B
- 1 mark for calculating the correct magnitude of the force
- 1 mark for the correct direction of the force

Question 2a.

Worked solution

The total mechanical energy (E_{total}) of carriage C1 is the sum of its kinetic energy at P and its gravitational potential energy; that is:

$$E_{total} = E_k + E_g = \frac{1}{2}mv^2 + mgh = \frac{1}{2} \times 850 \times 3.5^2 + 850 \times 98 \times 9.5 = 84\,341 \text{ J}$$

Since there is no energy loss due to friction, E_{total} remains the same at point Q; however, there is no gravitational potential energy at this point, so $E_{total} = E_k$.

Solving for v in $E_k = \frac{1}{2}mv^2 = 84\,341$ J, we obtain $v = 14.1$ m s^{-1}.

Mark allocation: 3 marks

- 1 mark for correctly calculating E_{total}
- 1 mark for identifying that $E_{total} = E_k$ at point Q
- 1 mark for the correct answer

Question 2b.

Worked solution

An isolated collision is modelled as one in which no net external forces, such as friction or air resistance, act. The only forces causing an effect on the momentum of the bodies in the system are those applied by other bodies in the system. Hence, the momentum of the system is conserved.

Mark allocation: 3 marks

- 1 mark for noting that no net external forces act
- 1 mark for noting that the only forces causing an effect on the momentum are those applied by interacting bodies within the system
- 1 mark for highlighting the outcome that the momentum of the system is conserved

TIPS

» Although gravity still acts on bodies in the system, it is cancelled by the normal forces in the system; hence, the system may still be considered isolated.

» It is also vital to distinguish the term 'isolated collision' from the term 'elastic collision'; elastic collision refers to the conservation of kinetic energy in the system. It is possible to conserve momentum without conserving kinetic energy.

Question 2c.

Worked solution

Since momentum is conserved, $p_{total,\,before} = p_{total,\,after} \Rightarrow (p_1 + p_2)_{after}$

As C2 is stationary before the collision, and both carriages move away at a common speed after the collision, the equation above simplifies to $m_1 u_1 = (m_1 + m_2)\,v_{common}$.

Substituting in the values gives $850 \times 14.1 = (850 + 550)\,v_{common}$, and solving for the common speed gives $v_{common} = 8.56$ m s^{-1}.

Mark allocation: 2 marks

- 1 mark for identifying conservation of momentum and deriving the equation involving the common speed after the collision

- 1 mark for the correct answer

Note: It is possible to award consequential marks for an answer based on a student's value obtained in **part a.**

$v_{common} = 0.607 \times$ Answer(a)

Question 2d.

Worked solution

The E_k of the combined carriages is given by $E_k = \dfrac{1}{2}mv^2 = \dfrac{1}{2} \times 1400 \times 8.56^2 = 51\,290$ J.

As the carriages stop at point R, all of the E_k has converted to E_g, meaning $E_g = mgh = 51\,290$ J.

Substituting in the values gives $1400 \times 9.8 \times h = 51\,290$, and solving for the height gives $h = 3.74$ m.

Mark allocation: 3 marks

- 1 mark for obtaining the E_k of the combined carriages

- 1 mark for equating the E_k to the E_g at point R

- 1 mark for the correct answer for the height

Note: It is possible to award consequential marks for an answer based on a student's value obtained in **part c.**

Thus, $E_k = 700 \times$ (Answer(c))^2 and $h = 0.05102 \times$ (Answer(c))^2.

 TIP

» Although the collision is isolated (i.e. the momentum is conserved), it is not elastic in that kinetic energy is not conserved. Therefore, the total mechanical energy of the system has changed, and it would be erroneous to use the total mechanical energy at P and equate it to the total mechanical energy at R.

Question 3a.

Worked solution

The friction force is the centripetal force on the parcel, which is given by

$$F_c = \frac{mv^2}{r} = \frac{4.5 \times 17^2}{8} = 163 \text{ N}.$$

Mark allocation: 2 marks

- 1 mark for correctly substituting values into the correct formula
- 1 mark for the correct answer

Question 3b.

Worked solution

Applying the same friction force, which provides the centripetal force on the parcel, we obtain a new equation for the new speed, $F_c = \frac{mv_2^2}{r^2} = 163 \text{ N}$.

Thus, the minimum radius would be $r_2 = \frac{mv_2^2}{r^2} = \frac{4.5 \times 19^2}{163} = 10 \text{ m}.$

Mark allocation: 3 marks

- 1 mark for equating the friction force to the new centripetal force equation
- 1 mark for correctly substituting values into the correct formula
- 1 mark for the correct answer

Note: It is possible to award consequential marks for a student using the value of friction force obtained in **part a.**

$$r_2 = \frac{1624.5}{\text{Answer(a)}}$$

Question 4

Worked solution

Teng is correct (or Sophie is incorrect).

As the mass drops downwards, it accelerates and gains kinetic energy. At the halfway point, the mass stops accelerating downwards as the spring force equals the weight force of the mass; thus the mass stops gaining kinetic energy. The mass will also start to slow down due to the increasing spring force; thus the mass starts losing kinetic energy. Hence the kinetic energy of the system reaches its maximum at the halfway point. Therefore Teng is correct.

At the halfway point, the gravitational potential energy of the mass is halved, with the half lost having been converted into both the kinetic energy of the moving mass and the elastic potential energy of the spring. Hence the majority of the total energy of the system is still *potential* energy. Therefore Sophie is incorrect.

Mark allocation: 3 marks

- 1 mark for stating that Teng is correct
- 1 mark for correctly explaining why Teng is correct
- 1 mark for correctly explaining why Sophie is incorrect

SET 6

Question 1a.

Worked solution

$$E_k = E_g$$
$$= mg\Delta h$$
$$= 200 \times 9.81 \times 2.5$$
$$= 4.9 \times 10^3 \text{ J}$$

Mark allocation: 2 marks

- 1 mark for $E_k = E_g$
- 1 mark for the correct answer

 TIP

> » VCAA generally asks students to use g = 9.81 m s^{-2}. In previous years the value has been 10 m s^{-2}. You should get into the habit of using the more exact value.

Question 1b.

Worked solution

$$m_1 v_1 = m_2 v_2$$
$$200 \times 20 = 285 \times v_2$$
$$v_2 = \frac{4000}{285}$$
$$= 14 \text{ m s}^{-1}$$

Mark allocation: 2 marks

- 1 mark for a correct statement of conservation of momentum (e.g. $m_1 v_1 = m_2 v_2$)
- 1 mark for the correct answer

Question 1c.

Worked solution

In an elastic collision, kinetic energy is conserved:

$$E_{k(initial)} = \frac{1}{2}mv^2$$

$$= \frac{1}{2} \times 200 \times 20^2$$

$$= 40\,000 \text{ J}$$

$$E_{k(final)} = \frac{1}{2} \times 285 \times 14^2$$

$$= 27\,930 \text{ J}$$

As kinetic energy is not conserved, the collision is inelastic.

Mark allocation: 2 marks

- 1 mark for an attempt to calculate kinetic energy before and after the collision
- 1 mark for the correct values of kinetic energy

Note: Consequential marks awarded for $E_{k(final)} = \frac{1}{2} \times 285 \times$ (Answer from **part b.**)^2.

TIP

» Remember that momentum is conserved in all collisions, but kinetic energy is conserved only in elastic collisions.

Question 1d.

Worked solution

The lost gravitational potential energy was used to break the wood.

$$mg\Delta h_1 = E_{wood} + mg\Delta h_2$$

$$200 \times 9.8 \times 5 = E_{wood} + 200 \times 9.8 \times 1.3$$

$$E_{wood} = 7252 \text{ J}$$

Mark allocation: 2 marks

- 1 mark for a correct expression of energy conservation
- 1 mark for the correct answer

TIP

» A good approach to energy conservation questions is to set out an equation that shows energy before the collision = energy after the collision.

Question 2

Worked solution

First, find the angle at the top of the triangle, between the string and the pole:

$$\cos \theta = \frac{12}{15}$$

$$\theta = \cos^{-1}\left(\frac{12}{15}\right)$$

$$= 36.9°$$

(Alternatively, the angle between the string and the horizontal may be calculated, which gives $\theta = 53.1°$.)

Next, draw a triangle to represent the forces that act on the ball.

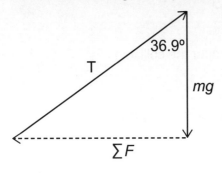

Finally, solve to find the net force:

$$\tan 36.9° = \frac{\Sigma F}{mg}$$

$$\Sigma F = mg \tan 36.9°$$

$$= 0.063 \times 9.8 \times \tan 36.9°$$

$$= 0.46 \text{ N}$$

OR

$$\tan 53.1° = \frac{mg}{\Sigma F}$$

$$\Sigma F = \frac{mg}{\tan 53.1°}$$

$$\Sigma F = \frac{0.063 \times 9.8}{\tan 53.1°}$$

$$= 0.46 \text{ N}$$

Mark allocation: 3 marks

- 1 mark for correctly calculating the angle
- 1 mark for correctly substituting values into the tan θ equation
- 1 mark for the correct answer

 TIP

» When using triangles like these, make sure you don't combine force quantities (i.e. tension, weight etc.) with dimension quantities (i.e. length, radius etc.).

Question 3a.

Worked solution

In the vertical direction:

$u = 0$ m s^{-1}

$a = -9.8$ m s^{-2}

$s = -2$ m

$t = ?$

$s = ut + \dfrac{1}{2}at^2$

$-2 = 0 \times t + \dfrac{1}{2} \times -9.8 \times t^2$

$t = \sqrt{\dfrac{2}{4.9}}$

$= 0.639$ s

In the horizontal direction:

$s = 5.0$ m

$t = 0.639$ s

$s = ut$

$5.0 = u \times 0.639$

$u = \dfrac{5.0}{0.639}$

$= 7.8$ m s^{-1}

Mark allocation: 3 marks

- 1 mark for the correct flight time
- 1 mark for correctly substituting the time into the distance formula
- 1 mark for the correct answer

Question 3b.

Worked solution

$u = 4.9$ m s^{-1}

$a = -9.8$ m s^{-2}

$s = -6$ m

$v = ?$

$v^2 = u^2 + 2as$

$\quad = 4.9^2 + 2 \times -9.8 \times -6$

$v = \sqrt{4.9^2 + 2 \times -9.8 \times -6}$

$\quad = 11.9$ m s^{-1}

Mark allocation: 2 marks

- 1 mark for substituting correctly
- 1 mark for the correct answer

Unit **3** | Area of Study **2**   How do things move without contact?

SET 1

Question 1

Answer: **D**

Explanatory notes

Magnetic field lines run out of a north pole and into a south pole.

Question 2

Answer: **B**

Explanatory notes

Use the right-hand push rule – fingers point in the direction of the magnetic field lines (right), thumb points in the direction of the current (into the page), palm faces the direction of the force.

Question 3

Answer: **A**

Explanatory notes

$$F = \frac{kq_1q_2}{r^2}$$
$$= \frac{8.99 \times 10^9 \times 0.01 \times 10^{-3} \times 0.03 \times 10^{-3}}{0.6^2}$$
$$= 7.5 \text{ N}$$

Like charges repel.

Question 4

Answer: **A**

Explanatory notes

$$W = qV$$
$$= 1.60 \times 10^{-19} \times 5000$$
$$= 8.00 \times 10^{-16} \text{ J}$$

Question 5

*Answer: **D***

Explanatory notes

Gravitational fields can attract only, never repel, so Y and Z cannot be gravitational fields.

Question 6

*Answer: **A***

Explanatory notes

Considering only the vertical motion of the electrons, the required acceleration may be obtained from $s = ut + \frac{1}{2}at^2$. Since the initial vertical velocity is zero, the equation simplifies to $s = \frac{1}{2}at^2$, so $a = \frac{2s}{t^2} = \frac{2 \times 39 \times 10^{-2}}{(2.3 \times 10^{-4})^2} = 1.47 \times 10^7\,\text{m s}^{-2}$.

Question 7

*Answer: **B***

Explanatory notes

Option A is incorrect because gravity alone is insufficient. Options C and D are incorrect because the magnetic field exerts a constant force perpendicular to the path of the electrons, and so the electrons will take a circular path. Option B is correct because only an electric field can cause the required constant vertical acceleration.

SET 2

Question 1

Answer: **D**

Explanatory notes

An electron has electric charge and therefore could be deflected by an electric field. It also has mass and therefore could be deflected by a gravitational field. Finally, a moving electron is a moving charge and so may be deflected by a magnetic field.

Question 2

Answer: **C**

Explanatory notes

Gravitational field strength is inversely proportional to the square of the distance between the centre of Duke and the centre of the planet Zorb; that is, $g \propto \dfrac{1}{r^2}$.

In the first instance, at an altitude of r above the surface of the planet, Duke is $2r_z$ from the centre of Zorb. This situation may be represented as $g_1 = \dfrac{k}{(2r_z)^2} = \dfrac{k}{4r_z^2}$.

When Duke's altitude increases to $2r$, he is $3r$ from the centre of Zorb.

This situation may be represented as $g_2 = \dfrac{k}{(3r_z)^2} = \dfrac{k}{9r_z^2}$.

Using the ratio $\dfrac{g_1}{g_2}$, the constants k and r can be eliminated and we end up with $\dfrac{g_1}{g_2} = \dfrac{9}{4}$.

As $g_1 = 9$ N kg^{-1}, then $g_2 = 4$ N kg^{-1}.

 TIP

» In questions involving a body orbiting a central body, it is vital to distinguish between the altitude of the orbiting body above the surface of the central body and the radial distance between the centres of masses.

Question 3

Answer: **D**

Explanatory notes

Depending on whether a magnetic body is present or not, and the magnetic polarity of any magnetic body that is significantly present within the magnetic field, the field may exert either no force (no magnetic body present), an attractive force (south pole significant within the field), or a repulsive force (north pole significant within the field).

Question 4

Answer: **A**

Explanatory notes

A static field is one whose field strength does not change with time, whereas a uniform field is one whose field strength does not change across a defined space. The gravitational field around a point mass fits the description of static and non-uniform because the field strength decreases as the square of the distance from the mass. The electric field between two plates at a constant potential difference is static because of the constant potential difference, but it is uniform within the confines of the two plates. The magnetic field around a solenoid connected to an AC power supply varies in time due to the alternating current, and therefore it is not static. The same could be said for the electric field between two plates connected to an AC power supply.

Question 5

Answer: **B**

Explanatory notes

Using Fleming's left-hand rule, the direction of the force on the current flowing from east to west, exerted by a magnetic field running south to north, is down.

SET 3

Question 1

Answer: **B**

Explanatory notes

The uniform electric field exerts a constant force in the downward direction, resulting in a parabolic path. This parabolic path is like that of projectile motion due to the constant, downward gravitational force.

The uniform magnetic field exerts a constant force perpendicular to the path of the particle, resulting in a circular path. This circular path is like that of uniform circular motion due to a centripetal force that acts perpendicular to the direction of motion.

Question 2

Answer: **B**

Explanatory notes

The magnitude of the force between an alpha particle and an electron is given by

$$F = \frac{kq_1q_2}{r^2} = \frac{8.99 \times 10^9 \times 2 \times 1.6 \times 10^{-19} \times 1.6 \times 10^{-19}}{(5 \times 10^{-6})^2} = 1.84 \times 10^{-17} \text{ N.}$$

Question 3

Answer: **C**

Explanatory notes

Due to the inverse square law, the electric field around an electric charge is not uniform in space. The magnitude of the force does not change in time, that is, it is constant. Since the particle is stationary in deep space, the electric field is also static.

Question 4

Answer: **C**

Explanatory notes

The magnitude of the magnetic force on a current-carrying wire is given by $F = nIlB$.

Hence, the magnetic field strength is $B = \dfrac{F}{nIl} = \dfrac{3.7 \times 10^{-3}}{10 \times 0.70 \times 0.45} = 1.17 \times 10^{-3} \text{ T.}$

Question 5

Answer: **B**

Explanatory notes

The gravitational potential is given by $g = \dfrac{GM_E}{R_E^2} = \dfrac{6.67 \times 10^{-11} \times 5.97 \times 10^{24}}{(6.38 \times 10^6)^2} = 9.78 \text{ N kg}^{-1}.$

SET 4

Question 1

Answer: **B**

Explanatory notes

Since both satellites are at the same altitude, they share a common orbital period. Hence, they must be orbiting at the same speed. It is not necessary for them to share the same orbital path. Their orbital period does not depend on their mass. Their relative masses do not affect their relative speeds.

TIP

» In satellite motion, the orbital period is independent of the mass of the satellite. The orbital period and the orbital radius are connected via the satellite equation, $\dfrac{R^3}{T^2} = \dfrac{GM_{central}}{4\pi^2}$.

Question 2

Answer: **B**

Explanatory notes

Option A applies only to the field generated by a point charge. Option C applies only to the field generated between a pair of plates connected to a DC supply. Option D is incorrect.

Question 3

Answer: **C**

Explanatory notes

The amount of charge carried may be determined from Coulomb's law (given on the VCAA formula sheet) $F = \dfrac{kq_1q_2}{r^2}$; thus $q_2 = \dfrac{Fr^2}{kq_1} = \dfrac{4.8 \times 10^{-3} \times (1.5 \times 10^{-6})^2}{8.99 \times 10^9 \times 1.60 \times 10^{-19}} = 7.5 \times 10^{-6}$ C.

TIP

» A mistake that students often make is forgetting to square quantities in formulas, such as the distance between the two charges when calculating the magnitude of the force. Other formulas involving squares (or cubes) include $F = \dfrac{GM_1M_2}{r^2}$ and $F_c = \dfrac{mv^2}{r}$.

Question 4

Answer: **A**

Explanatory notes

The direction of a magnetic field is determined by the force on an imaginary north monopole. A north monopole at P will be repelled by the north pole of the magnet on the left, and will be attracted by the south pole of the magnet on the right, both with equal magnitude; thus, the vector addition of these two forces will be a force directed to the right, as shown in the diagram below.

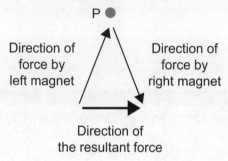

Question 5

Answer: **B**

Explanatory notes

The planets may be considered as satellites of the star Gliese 357. Thus, the satellite equation (derived below) may be applied to their orbital periods and radii; that is, $\frac{R^3}{T^2}$ is constant for all satellites of the star. The constant may be calculated using the data

for GJ 357 b: $\frac{R^3}{T^2} = \frac{(5 \times 10^6)^3}{3.9^2} = 8.218 \times 10^{18}$.

The orbital radius of GJ 357 d can then be calculated using

$R = \sqrt[3]{8.22 \times 10^{18} \times T^2} = \sqrt[3]{8.22 \times 10^{18} \times 56^2} = 2.95 \times 10^7$ km.

The satellite equation is derived by equating the gravitational force of the central body on the satellite and the centripetal force on the satellite, as the satellite's motion may be considered a uniform circular motion: $F_g = F_c \rightarrow \frac{GM_c m}{R^2} = \frac{m 4\pi^2 R}{T^2}$.

Rearranging the equation, we obtain $\frac{R^3}{T^2} = \frac{GM_c}{4\pi^2}$, which is constant for all satellites of the central mass.

 TIP

» Some formulas, such as the one for the satellite equation, are not provided in the VCAA formula sheet. Therefore, it would be advantageous for you to include this formula (and other useful ones) in the A3 pre-written notes that you are permitted to bring to the Physics examination.

SET 5

Question 1a.

Worked solution

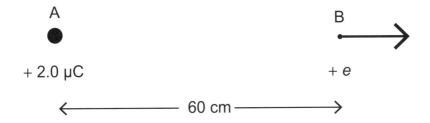

A

+ 2.0 µC

B

+ e

←————— 60 cm —————→

Mark allocation: 1 mark

- 1 mark for an arrow from point B, horizontally to the right

Question 1b.

Worked solution

$$E = \frac{kq}{r^2}$$

$$= \frac{8.99 \times 10^9 \times 2.0 \times 10^{-6}}{0.6^2}$$

$$= 5.0 \times 10^4 \text{ N C}^{-1}$$

Mark allocation: 2 marks

- 1 mark for substituting values correctly

- 1 mark for the correct answer

 TIP

» A very common mistake is forgetting to square the radius when entering values into a calculator. Also, ensure that you check the values before entering them into your calculator.

Question 2a.

Worked solution

$$F = qvB$$

$$= 1.60 \times 10^{-19} \times 1.5 \times 10^7 \times 8.0 \times 10^{-3}$$

$$= 1.92 \times 10^{-14} \text{ N}$$

Mark allocation: 2 marks

- 1 mark for substituting values correctly

- 1 mark for the correct answer

Question 2b.

Worked solution

Direction C – down the page.

Mark allocation: 1 mark

- 1 mark for the correct answer

Note: Use the right-hand slap rule – the thumb points in the direction of the current (opposite to the direction of electron motion) and the fingers point in the direction of the magnetic field. The palm points in the direction of the force.

Question 2c.

Worked solution

$$r = \frac{mv}{qB}$$

$$= \frac{9.11 \times 10^{-31} \times 1.5 \times 10^{7}}{1.60 \times 10^{-19} \times 8.0 \times 10^{-3}}$$

$$= 0.011 \text{ m}$$

Mark allocation: 2 marks

- 1 mark for substituting values correctly
- 1 mark for the correct answer

Question 3a.

Worked solution

Use $g = \dfrac{GM}{r^2} = \dfrac{6.67 \times 10^{-11} \times 2.9 \times 10^{30}}{(1.4 \times 10^{12})^2} = 9.9 \times 10^{-5}$ N kg^{-1}.

Mark allocation: 2 marks

- 1 mark for correctly substituting values into the correct formula
- 1 mark for the correct answer

Question 3b.

Worked solution

The acceleration of the planet is the same as the gravitational field strength at the orbital radius of the planet; that is, 9.9×10^{-5} m s^{-2}.

Mark allocation: 1 mark

- 1 mark for the correct answer

Question 3c.

Worked solution

No. Referring to the satellite equation, $\dfrac{R^3}{T^2} = \dfrac{GM_{central}}{4\pi^2}$, the orbital radius and period are related to the mass of the central body, which is the star. Hence, the mass of the planet is not involved and, therefore, cannot be determined.

Mark allocation: 3 marks

- 1 mark for answering 'no'
- 1 mark for using an appropriate equation, such as the satellite equation or $\dfrac{GMm}{r^2} = mg$
- 1 mark for an appropriate explanation that the mass of the planet does not figure in the equation and so cannot be found

Note: It is also acceptable to explain that since $a = g$, it is not possible to determine the mass of the planet. The planet is in freefall about the star just as an apple is in freefall when it falls from a tree, and you cannot determine the mass of an apple from its acceleration.

SET 6

Question 1a.

Worked solution

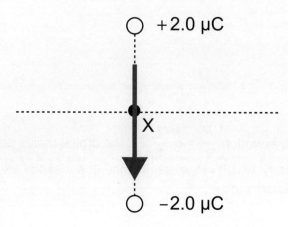

Mark allocation: 1 mark

- 1 mark for an arrow pointing down

Note: The direction of the electric field is the same as the direction of the force the field exerts on a small positive charge.

Question 1b.

Worked solution

direction A

Mark allocation: 1 mark

- 1 mark for the correct answer

Note: The negative charge would initially move in the opposite direction to the electric field vector.

Question 1c.

Worked solution

Use $F = \dfrac{kq_1 q_2}{r^2} = \dfrac{8.99 \times 10^9 \times 2.0 \times 10^{-6} \times 2.0 \times 10^{-6}}{(0.50)^2} = 0.14\ \text{N}.$

Mark allocation: 2 marks

- 1 mark for correctly substituting values into the correct formula
- 1 mark for the correct answer

Question 2

Worked solution

Use $F = IlB = 0.6 \times 0.12 \times 0.8 = 0.0576$ N.

Using the right-hand slap rule or Fleming's left-hand rule, the direction of the force should be down (direction D).

Mark allocation: 3 marks

- 1 mark for correctly substituting values into the correct formula to obtain the force magnitude
- 1 mark for the correct answer
- 1 mark for correctly identifying the direction

Question 3a.

Worked solution

Clockwise: the current flows in the direction D → C → B → A.

The force on side DC is down and the force on side AB is up, according to the right-hand slap rule.

Mark allocation: 3 marks

- 1 mark for the correct direction
- 1 mark for correctly identifying the direction of current flow through at least one side of the loop
- 1 mark for correctly identifying the direction of the force on at least one side of the loop

Question 3b.

Worked solution

The direction of the current flowing in the coil will reverse. This ensures that the coil continues to rotate in the same direction.

Mark allocation: 2 marks

- 1 mark for identifying the reversal of current flow
- 1 mark for identifying continuous rotation

SET 7

Question 1a.

Worked solution

$$F = \frac{GM_1M_2}{r^2}$$

$$= \frac{6.67 \times 10^{-11} \times 410 \times 1.3 \times 10^{22}}{(1.25 \times 10^7 + 1.2 \times 10^6)^2}$$

$$= 1.9 \text{ N}$$

Mark allocation: 3 marks

- 1 mark for the correct calculation of the radius ($12\,500\,000$ m + $1\,200\,000$ m = 1.37×10^7 m)
- 1 mark for the correct calculation of the force (award this mark even if altitude is used instead of radius; e.g. $F = 2.3$ N if $r = 1.25 \times 10^7$ m)
- 1 mark for using 2 significant figures in the answer

TIP

» Remember that orbital radius, not altitude, is used in calculations such as this.

Question 1b.

Worked solution

Find the area under the graph (12 squares) and multiply by the mass of the probe:

$$12 \times 2.5 \times 10^6 \times 1.0 \times 10^{-3} = 30\,000 \text{ J kg}^{-1}$$

$$30\,000 \times 410 = 1.23 \times 10^7 \text{ J}$$

Mark allocation: 3 marks

- 1 mark for attempting to find the area under the graph (even if incorrect)
- 1 mark for calculating the correct area under the graph (accept 11 to 12 squares, $28\,750 \pm 1250$ J kg^{-1})
- 1 mark for finding the correct energy (accept from 1.1×10^7 J to 1.3×10^7 J)

TIP

» The following guidelines may help you to count squares more quickly – if more than half a square is filled, count it as 1. If less than half a square is filled, count it as zero.

Question 2a.

Worked solution

Use $E = \dfrac{v}{d} = \dfrac{110}{0.12} = 9.2 \times 10^2$ V m^{-1}.

Mark allocation: 2 marks

- 1 mark for correctly substituting values into the correct formula
- 1 mark for the correct answer

Question 2b.

Worked solution

Use work done, $W = Vq = \Delta E_k$. Since the initial kinetic energy is zero,

$Vq = \dfrac{1}{2}mv^2 \Rightarrow 110 \times 1.60 \times 10^{-19} = \dfrac{1}{2} \times 9.11 \times 10^{-31} \times v^2$.

Solve for v to obtain $v = 6.22 \times 10^6$ m s^{-1}.

Mark allocation: 3 marks

- 1 mark for relating the work done by the electric field on the electron to the change in kinetic energy
- 1 mark for correctly substituting values into the formula of that relationship
- 1 mark for the correct answer

Question 2c.

Worked solution

The electrons in region Z require a force directed up the page. The current associated with the flow of electrons is directed to the left. Using the right-hand slap rule or Fleming's left-hand rule, the magnetic field should be directed out of the page.

Mark allocation: 3 marks

- 1 mark for identifying the correct direction of the force
- 1 mark for identifying the correct direction of the current
- 1 mark for identifying the correct direction of the magnetic field

 TIP

» **The direction of conventional current is opposite to the direction of the flow of electrons, which are negative charges. Electric current may be thought of as a flow of positive charges.**

SET 8

Question 1a.

Worked solution

0 N

Mark allocation: 1 mark

- 1 mark for correctly identifying that the electrical force on a body without any electrical charge is 0 N

Question 1b.

Worked solution

4.0×10^{-3} mC

The positive charges on S1 will be distributed evenly between the two identical spheres, so eventually the charge on both spheres will be equal to half that which was originally on S1. The answer must be to 2 significant figures and use the scale of 10 mC = 10^{-3} µC.

Mark allocation: 2 marks

- 1 mark for the correct answer
- 1 mark for giving the answer to 2 significant figures

TIP

» It is easy to confuse significant figures and decimal places. To avoid this, express your numerical answers in standard index form: $m \times 10^n$, where $1 \le m < 10$. The number of significant figures is determined by the number of significant figures in the value of m. For example, the number 0.0078 has 4 decimal places, but when expressed as 7.8×10^{-3} it becomes clear that the number has only 2 significant figures.

Question 1c.

Worked solution

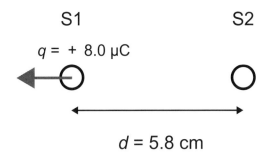

Since both spheres are positively charged, the spheres would repel each other. S1 would experience a force to the left by S2.

Mark allocation: 2 marks

- 1 mark for an arrow pointing left
- 1 mark for the correct explanation of answer

Question 2a.

Worked solution

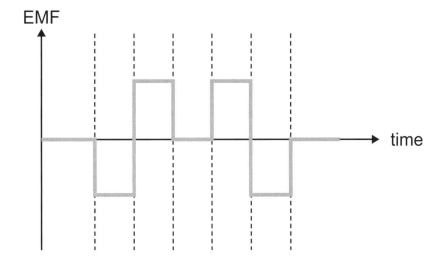

Explanatory notes

An EMF is induced in the loop due to changes in the magnetic field strength of the electromagnet, according to the formula $\varepsilon = -N \dfrac{\Delta \phi}{\Delta t}$, with the magnitude of the EMF proportional to the rate of change of the field strength.

Mark allocation: 3 marks

- 1 mark for four waves of equal magnitude drawn in the correct time intervals – if incorrect, no further marks are to be awarded
- 1 mark for square waves, not sinusoidal or any other shape
- 1 mark for a symmetrical pattern of 'down-up-up-down' or 'up-down-down-up'

Question 2b.

Worked solution

8.22×10^{-5} Wb

The magnetic flux through the loop is given by $\phi = BA = 0.06 \times 13.7 \times 10^{-4} = 8.22 \times 10^{-5}$ Wb.

Mark allocation: 1 mark

- 1 mark for the correct answer

Question 2c.

Worked solution

9.13×10^{-4} V

The EMF generated is given by $\varepsilon = -N\dfrac{\Delta\phi}{\Delta t} = \dfrac{8.22 \times 10^{-5}}{0.09} = 9.13 \times 10^{-4}$ V.

Mark allocation: 2 marks

- 1 mark for substituting correct values into the correct formula
- 1 mark for the correct answer

Note: It is possible to award consequential marks for using the value of the flux obtained in **part b.** EMF = (Answer **part b.**)/0.09.

Question 3a.

*Answer: **E***

Worked solution

Using the right-hand push rule or Fleming's left-hand rule, the direction of the field is upwards from north to south, the direction of the current is from positive to negative through the conductor, so the direction of force is away from the magnet towards the current source.

Mark allocation: 2 marks

- 1 mark for the correct direction
- 1 mark for the correct explanation of the direction

Question 3b.

Worked solution

2.88×10^{-4} T

Using $F = nIlB \rightarrow B = \dfrac{F}{nIl} = \dfrac{7.3 \times 10^{-6}}{0.39 \times 6.5 \times 10^{-2}} = 2.88 \times 10^{-4}$ T.

Mark allocation: 2 marks

- 1 mark for correctly substituting values into the formula
- 1 mark for the correct answer

SET 9

Question 1a.

Worked solution

7.14×10^{10} kg

The orbital radius is half the orbital diameter: $R = \dfrac{3480}{2} = 1740$ m. The mass of the asteroid, which is the central body in the satellite equation, can be found from

$$\frac{R^3}{T^2} = \frac{GM_{central}}{4\pi^2} \rightarrow M_{central} = \frac{4\pi^2 R^3}{GT^2} = \frac{4\pi^2 \times 3480^3}{6.67 \times 10^{-11} \times 209\,000^2} = 7.14 \times 10^{10} \text{ kg}.$$

Mark allocation: 3 marks

- 1 mark for correctly stating the satellite equation
- 1 mark for correctly substituting the values into the equation
- 1 mark for the correct answer

Question 1b.

Worked solution

$g = 5.29 \times 10^{-5}$ N kg^{-1}

The gravitational field strength is given by

$$g = \frac{GM_{central}}{r^2} = \frac{6.67 \times 10^{-11} \times 7.14 \times 10^{10}}{300^2} = 5.29 \times 10^{-5} \text{ N kg}^{-1}.$$

Mark allocation: 2 marks

- 1 mark for correctly substituting values into the equation
- 1 mark for the correct answer

Note: It is possible to award consequential marks for using the value of the mass obtained in **part a.** $g = 7.41 \times 10^{-16} \times$ (Answer **part a.**) N kg^{-1}.

Question 2a.

Worked solution

Equating the electrical potential energy of the electrons due to the charged plates and the kinetic energy of the electrons gives $E_p = E_k \rightarrow qV = \dfrac{1}{2}mv^2$.

After rearranging for velocity, we get $v = \sqrt{\dfrac{2qV}{m}}$.

Substituting the values gives $v = \sqrt{\dfrac{2 \times 1.60 \times 10^{-19} \times 2.4 \times 10^6}{9.11 \times 10^{-31}}} = 9.18 \times 10^8 \text{ m s}^{-1}$.

Mark allocation: 2 marks

- 1 mark for correctly equating the E_p and the E_k of the electrons
- 1 mark for substituting the values into the equation and demonstrating that the given value is obtained

Question 2b.

Worked solution

The expected velocity is higher than the speed of light, $c = 3.00 \times 10^8$ m s^{-1}. This is not possible because, according to Einstein's theory of special relativity, the speed of light is an absolute limit for all observers regardless of their frames of reference.

Mark allocation: 2 marks

- 1 mark for pointing out that the expected velocity is superluminal (i.e. faster than light)
- 1 mark for stating that this is not possible

Question 3a.

Worked solution

5.8×10^3 s

First, the orbital radius of RISAT-2BR1 must be calculated by summing the altitude of the satellite and the radius of Earth: 576 + 6370 = 6946 km.

The period is then determined from equating the centripetal acceleration of the satellite $a = \dfrac{4\pi^2 r}{T^2}$ and the gravitational field strength at the satellite's altitude $g = \dfrac{GM_E}{r^2}$, that is, $\dfrac{4\pi^2 r}{T^2} = \dfrac{GM_E}{r^2}$.

Thus, $T = \sqrt{\dfrac{4\pi^2 r^3}{GM_E}} = \sqrt{\dfrac{4\pi^2 \times (6.946 \times 10^6)^3}{6.67 \times 10^{-11} \times 5.97 \times 10^{24}}} = 5.76 \times 10^3$ s.

Mark allocation: 4 marks

- 1 mark for the correct answer for orbital radius
- 1 mark for correctly deriving the equation for the period

  Note: This mark may be awarded for providing the formula without deriving it.

- 1 mark for correctly substituting the values into the period equation
- 1 mark for the correct answer

Question 3b.

Worked solution

centripetal acceleration, orbital period, orbital speed

All these characteristics are dependent only on the orbital radius. The centripetal force on each satellite is different for each of them because they are the product of the satellite's mass and its centripetal acceleration.

Mark allocation: 2 marks

- 2 marks for all three correct answers
- 1 mark for any two correct answers

Note: Deduct 1 mark from the above marks if the wrong answer is also circled.

SET 10

Question 1a.

Worked solution

$$g = \frac{GM_E}{r^2} = \frac{6.67 \times 10^{-11} \times 5.97 \times 10^{24}}{(27 \times 10^6)^2} = 0.546 \text{ N kg}^{-1}$$

Mark allocation: 2 marks

- 1 mark for the correct formula
- 1 mark for correctly substituting the values for G, M_E and r

Note: As no marks are awarded for a correct final answer, the substituted values must be correct.

Question 1b.

Worked solution

$3.8 \times 10^3 \text{ m s}^{-1}$

Since the centripetal acceleration is equal to the gravitational field strength, that is, $a_c = \frac{v^2}{r} = g$, the formula can be rearranged to obtain $v = \sqrt{rg} = \sqrt{2.7 \times 10^7 \times 0.547} = 3.84 \times 10^3 \text{ m s}^{-1}$.

Mark allocation: 2 marks

- 1 mark for correctly substituting the values into the formula
- 1 mark for the correct answer

Question 1c.

Worked solution

$2.4 \times 10^9 \text{ J}$

The difference in the gravitational potential energy is the area under the graph of gravitational field strength vs orbital radius multiplied by the mass of the satellite. The area under the graph may be calculated using the trapezium area formula or similar:

$$A = \frac{1}{2}(a + b)h$$

$$= \frac{1}{2}(0.547 + 0.443)(30 \times 10^6 - 27 \times 10^6)$$

$$= 1.49 \times 10^6 \text{ J kg}^{-1}$$

This value is multiplied by the mass of the satellite, so $\Delta E_g = 1.49 \times 10^6 \times 1630 = 2.4 \times 10^9 \text{ J}$.

Mark allocation: 3 marks

- 1 mark for a correct formula to calculate the area under the graph
- 1 mark for the correct answer for the specific energy value of $1.49 \times 10^6 \text{ J kg}^{-1}$
- 1 mark for the correct answer for the energy value of $2.4 \times 10^9 \text{ J}$

Question 2a.i.

Worked solution

anticlockwise

Mark allocation: 1 mark

- 1 mark for the correct direction of rotation

Question 2a.ii.

Worked solution

According to the right-hand push rule, the magnetic field is to the right while the current from the positive terminal of the DC supply enters terminal P and flows from K to L. Thus, the direction of the force on side KL is down. This downward force on KL would cause the coil to rotate anticlockwise.

Mark allocation: 2 marks

- 1 mark for the correct explanation of the direction of the force on side KL, using the right-hand push rule
- 1 mark for connecting the direction of the force on KL to the direction of rotation

Question 2b.

Worked solution

9.9×10^{-4} N

First, the current flowing in the coil is determined from the data given, using Ohm's law:
$V = IR$.
$$I = \frac{V}{R} = \frac{9}{18} = 0.5 \text{ A}.$$
The magnetic force on side KL is then calculated using
$F = nBIl = 20 \times 4.5 \times 10^{-3} \times 0.5 \times 2.2 \times 10^{-2} = 9.90 \times 10^{-4}$ N.

Mark allocation: 3 marks

- 1 mark for the correct value of the current
- 1 mark for correctly substituting values into the magnetic force formula
- 1 mark for the correct answer

Question 2c.

Worked solution

The rectangular coil will oscillate/vibrate without turning. This is because the magnetic force on the sides perpendicular to the magnetic field will alternate or change direction constantly due to the AC supply.

Mark allocation: 2 marks

- 1 mark for stating that the rectangular coil will not turn
- 1 mark for correctly stating that the AC supply will cause the magnetic force on the sides to change direction constantly

SET 11

Question 1a.

Worked solution

The path of the electron is the shape of a parabola because it is experiencing a constant vertical downward force and acceleration while also travelling with a constant horizontal component of its velocity. The combination of these two motions results in a projectile motion that is parabolic in shape.

Mark allocation: 3 marks

- 1 mark for stating that the electron experiences vertical acceleration
- 1 mark for stating that the electron has a constant horizontal component of its velocity
- 1 mark for stating that the combination of these two motions results in a projectile motion or a parabolic path

Question 1b.

Worked solution

The field strength is calculated using $E = \dfrac{V}{d} = \dfrac{100}{0.4} = 250$ V m^{-1}. The alternative unit is N C^{-1}.

Mark allocation: 3 marks

- 1 mark for the correct formula
- 1 mark for correctly substituting the values for V and d
- 1 mark for the correct unit

Note: As no marks are awarded for a correct final answer, the substituted values must be correct.

Question 1c.

Worked solution

4.4×10^{13} m s^{-2}

The force on the electron by the uniform electric field is given by
$F = qE = 1.60 \times 10^{-19} \times 250 = 4.0 \times 10^{-17}$ N.

The acceleration of the electron is calculated from $F = ma$:
$a = \dfrac{F}{m} = \dfrac{4 \times 10^{-17}}{9.1 \times 10^{-31}} = 4.40 \times 10^{13}$ m s^{-2}

Mark allocation: 3 marks

- 1 mark for correctly substituting values into the formula for force on the electron
- 1 mark for calculating the force correctly
- 1 mark for calculating the acceleration correctly

Question 1d.

Worked solution

4.7×10^6 m s^{-1}

The transit time of the electron across the uniform electric field may be calculated from its horizontal velocity and the plate width:

$$t = \frac{w}{u} = \frac{0.25}{3.5 \times 10^6} = 7.14 \times 10^{-8} \text{ s}$$

This is also the time over which the electron experiences a downward acceleration due to the uniform electric field. Since the initial vertical velocity is zero, the final downward velocity of the electron is given by

$$v_v = u_v + at = at = 4.4 \times 10^{13} \times 7.14 \times 10^{-8} = 3.14 \times 10^6 \text{ m s}^{-1}$$

The speed of the electron as it exits the uniform electric field is the vector addition of the horizontal velocity and the vertical velocity, using Pythagoras' formula:

$$v = \sqrt{v_H^2 + v_V^2} = \sqrt{(3.5 \times 10^6)^2 + (3.14 \times 10^6)^2}$$
$$= 4.70 \times 10^6 \text{ m s}^{-1}$$

Mark allocation: 4 marks

- 1 mark for correctly calculating the transit time of the electron
- 1 mark for correctly calculating the final vertical velocity of the electron
- 1 mark for correctly substituting the values into Pythagoras' formula
- 1 mark for the correct answer

Note: It is possible to award a consequential mark for using the value of the acceleration obtained in **part c.** to calculate the final vertical velocity of the electron: v_v = (Answer **part c.**) $\times$ 7.14 $\times$ 10^{-8}. Further consequential marks may be awarded for calculating the exit speed of the electron.

Unit 3 | Area of Study 3 How are fields used in electricity generation?

SET 1

Question 1

Answer: A

Explanatory notes

The EMF is given by $\varepsilon = -N \dfrac{\Delta(BA)}{\Delta t}$. All the suggested actions increase the EMF **except** for increasing the period (Δt), which reduces the EMF.

Question 2

Answer: C

Explanatory notes

Since power loss in the transmission line is $P_{loss} = I^2 R$, lowering the current in the transmission line by stepping up the voltage will reduce the transmission loss.

Option A cannot be realised because there is always a resistance in the transmission line (apart from superconducting material).

Option B is incorrect, as transformers do not increase power and are modelled to have the same power output as the power input.

Option D is incorrect, as conservation of energy has nothing to do with the resistance of the transmission line.

Question 3

Answer: C

Explanatory notes

The turns ratio of the step-down transformer is given by $\dfrac{V_1}{V_2} = \dfrac{N_1}{N_2} = \dfrac{240}{15} = 16$.

Question 4

Answer: B

Explanatory notes

The RMS voltage of the transformer output is the DC equivalent. Hence, the DC value is the same as the AC value after the step-down transformer.

Question 5

Answer: **B**

Explanatory notes

The output voltage of an AC generator is given by $\varepsilon = -N\dfrac{\Delta\phi_B}{\Delta t}$, the EMF formula given on the VCAA formula sheet. Options A and C will double the output voltage but they will **also** change the period of the output. Option D will triple the output voltage.

Question 6

Answer: **C**

Explanatory notes

The use of a split-ring commutator will produce a DC voltage. Since the magnetic flux through the rotating coil is sinusoidal, the magnitude of the DC voltage will be fluctuating.

Question 7

Answer: **D**

Explanatory notes

The purpose of the inverter is to convert DC electricity to AC electricity for grid connection.

SET 2

Question 1a.

Worked solution

$\phi = BA$

$\quad = 6.0 \times 10^{-3} \times 0.03$

$\quad = 1.8 \times 10^{-4}$ Wb

Mark allocation: 2 marks

- 1 mark for the correct value for flux
- 1 mark for the correct unit

 TIP

> »   Calculations of flux do not include the number of loops, *n*.

Question 1b.

Worked solution

Find the time taken for one-quarter turn of the coil:

$$\Delta t = \frac{1}{4} \times \frac{1}{20}$$

$$= 0.0125 \text{ s}$$

Calculate EMF:

$$\varepsilon = N \frac{\Delta\phi}{\Delta t}$$

$$= 200 \times \frac{1.8 \times 10^{-4}}{0.0125}$$

$$= 2.9 \text{ V}$$

Mark allocation: 2 marks

- 1 mark for the correct Δt value
- 1 mark for the correct answer

Question 1c.

Worked solution

$$V_{peak} = V_{RMS} \times \sqrt{2}$$

$$= 3.5 \times \sqrt{2}$$

$$= 4.9 \text{ V}$$

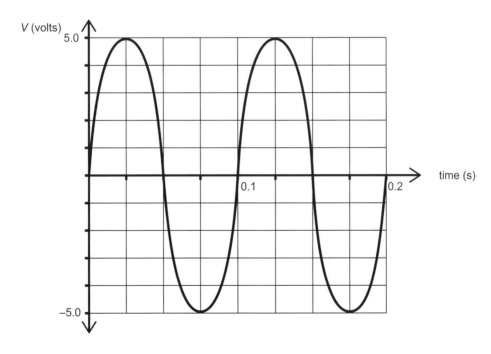

Note: The opposite phase (flipped graph) is equally correct.

Mark allocation: 2 marks

- 1 mark for a graph that peaks at ± 4.9 V (5 V is acceptable)
- 1 mark for any sinusoidal curve that shows alternating current and that has a period of 0.1 s

Question 2a.

Worked solution

$V_{drop} = I \times R$

$\qquad = 10 \times 6 \qquad\qquad V_{supply} = 250 - 60$

$\qquad = 60\ V \qquad\qquad\qquad\qquad\ \ = 190\ V$

Mark allocation: 2 marks

- 1 mark for the correct calculation of V_{drop}
- 1 mark for the correct calculation of V_{supply}

Question 2b.

Worked solution

$10 \div 5 = 2.0\ A$

Mark allocation: 1 mark

- 1 mark for the correct calculation of the current

Note: As the current passes from the transmission lines to the house, the 5:1 step-down transformer will step-up the current by a factor of 5. Therefore, the current in the transmission lines is 5 times lower than in the house.

Question 2c.

Worked solution

$P_{loss} = I^2 R$

$\qquad = 2^2 \times 6$

$\qquad = 24\ W$

$$\frac{24}{2500} \times 100 = 0.96\%$$

Mark allocation: 3 marks

- 1 mark for $P_{loss} = I^2 R$
- 1 mark for the correct power loss
- 1 mark for the correct percentage calculation

 TIP

» When power is lost in transmission lines, the current is not affected (recall Kirchhoff's current law). The resistance of the transmission lines causes a drop in electrical potential energy (voltage).

SET 3

Question 1a.

Worked solution

Use $V_{p\text{-}p} = 2V_{RMS}\sqrt{2} = 2 \times 240 \times \sqrt{2} = 679$ V.

Mark allocation: 1 mark

- 1 mark for the correct answer

Question 1b.

Worked solution

Rearrange the formula for power, $P = VI$, to obtain $I = \dfrac{P}{V} = \dfrac{8.4 \times 10^5}{240} = 3500$ A$_{RMS}$.

Mark allocation: 2 marks

- 1 mark for correctly substituting values into the correct formula, to obtain the current
- 1 mark for the correct answer

Note: Zero marks are to be awarded if working is not shown.

Question 1c.

Worked solution

A turns ratio of 100 would reduce the secondary current in T1 by the same ratio.
The transmission line current would then be $3500 \div 100 \Rightarrow I_{transmission} = 35$ A$_{RMS}$.

The transmission loss of the power line is $P_{loss} = I^2R = 35^2 \times 3.0 = 3675$ W.

Mark allocation: 3 marks

- 1 mark for obtaining the correct transmission line current
- 1 mark for correctly substituting values into the correct formula to obtain the power loss
- 1 mark for the correct answer

Question 2a.

Worked solution

Use $\varepsilon = -\dfrac{BA}{\Delta t} = -Blv = -0.8 \times 0.12 \times 0.15 = -0.0144$ V.

Mark allocation: 2 marks

- 1 mark for correctly substituting values into the correct formula to obtain the EMF magnitude
- 1 mark for the correct answer

 TIP

» Although a straight conductor is used here rather than the often-used closed loop, the conductor 'sweeps' out an area over a period of time, thereby giving the ratio of $\dfrac{A}{\Delta t}$ that is required in the equation. You need to look out for variations in the physical set-up of the questions, such as that given here, as part of your exam preparation.

Question 2b.

Worked solution

from P to Q

Mark allocation: 1 mark

- 1 mark for correctly identifying the direction

Note: Imagine a positive charge on the conductor between P and Q. Since the conductor moves upwards, that positive charge represents a current moving upwards in a magnetic field that is directed from left to right. That positive charge would experience a magnetic force on it. Using the right-hand push rule or Fleming's rule, the direction of the magnetic force is away from P towards Q.

Question 3a.

Worked solution

91 mT

The area of the circular loop must be calculated first: $A = \dfrac{\pi D^2}{4} = \dfrac{\pi(250 \times 10^{-3})^2}{4} = 4.91 \times 10^{-2}$ m^2

The EMF generated is given by $\varepsilon = -N\dfrac{\Delta\phi}{\Delta t} = -N = \dfrac{\Delta(BA)}{\Delta t}$.

Since the magnetic flux due to the car part changes linearly from zero to maximum,

the required magnetic field strength may be calculated from

$B = \dfrac{\varepsilon t}{NA} = \dfrac{55 \times 10^{-3} \times 0.65}{-8 \times 4.91 \times 10^{-2}} = 9.10 \times 10^{-2}$ T.

Mark allocation: 3 marks

- 1 mark for correctly calculating the area of the circular loop
- 1 mark for correctly substituting values into the formula for the magnetic field strength
- 1 mark for the correct answer in mT

Question 3b.

Worked solution

The production engineer could speed up the conveyor belt. This would decrease the time for the magnetic flux to change, resulting in a sufficient EMF that compensates for the lower magnetic field strength.

Mark allocation: 2 marks

- 1 mark for correctly suggesting that the conveyor belt could be sped up

- 1 mark for the correct explanation that the faster conveyor belt would reduce the time for the flux to change

Question 4

Worked solution

An inverter is necessary in a rooftop solar installation so that the direct current (DC) electricity produced by the photovoltaic cells can be converted to alternating current (AC) electricity, which is the standard form of electricity used in most homes and required for connection to the electrical grid.

Mark allocation: 2 marks

- 1 mark for identifying that solar installations produce DC, whereas households run on AC

- 1 mark for explaining that an inverter is required to convert DC to AC, allowing solar power to be used in the home

SET 4

Question 1a.

Worked solution

81.8 A

Using $P = IV \rightarrow I = \dfrac{P}{V} = \dfrac{4.5 \times 10^3}{55} = 81.8$ A.

Mark allocation: 2 marks

- 1 mark for correctly substituting values into the formula
- 1 mark for the correct answer

Question 1b.

Worked solution

24.5 V

$V_{drop} = IR = 81.8 \times 0.3 = 24.5$ V.

Mark allocation: 2 marks

- 1 mark for correctly substituting values into the formula
- 1 mark for the correct answer

Note: It is possible to award consequential marks for using the value of the current obtained in **part a.**: $V_{drop} = IR =$ (Answer **part a.**) $\times 0.3$ V.

Question 1c.

Worked solution

30.5 V

$V_{lamp} = V_{generator} - V_{drop} = 55 - 24.5 = 30.5$ V

Mark allocation: 2 marks

- 1 mark for correctly substituting values into the formula
- 1 mark for the correct answer

Note: It is possible to award consequential marks for using the value of the voltage drop obtained in **part b.**: $V_{lamp} = V_{generator} -$ (Answer **part b.**) V.

Question 1d.i.

Worked solution

27.3 A

$I_{line} = I_{sec} = \dfrac{I_{prim}}{N} = \dfrac{81.8}{3} = 27.3$ A

Mark allocation: 1 mark

- 1 mark for the correct answer

Note: It is possible to award consequential marks for using the value of the current obtained in

part a.: $I_{line} = \dfrac{(\text{Answer } \textbf{part a.})}{3}$.

Question 1d.ii.

Worked solution

165 V

$V_{sec} = V_{prim} N = 55 \times 3 = 165$ V

Mark allocation: 1 mark

- 1 mark for the correct answer

Question 1e.

Worked solution

8.18 V

Using the new value for the transmission line current, $V_{drop} = IR = 27.3 \times 0.3 = 8.18$ V.

Mark allocation: 2 marks

- 1 mark for correctly substituting values into the formula
- 1 mark for the correct answer

Note: It is possible to award consequential marks for using the value of the current obtained in **part d.i.**: $V_{drop} = IR = $ (Answer **part d.i.**) $\times 0.3$ V.

Question 1f.

Worked solution

157 V

$V_{prim\ step-down} = V_{sec\ step-up} - V_{drop} = 165 - 8.18 = 156.8$ V

Mark allocation: 2 marks

- 1 mark for correctly substituting values into the formula
- 1 mark for the correct answer

Note: It is possible to award consequential marks for using the value of the voltage obtained in **part d.ii.** and the voltage drop obtained in **part e.**

$V_{prim\ step-down} = $ (Answer **part d.ii.**) $-$ (Answer **part e.**) V

Question 1g.

Worked solution

52.3 V

$V_{sec} = \dfrac{V_{prim}}{N} = \dfrac{156.8}{3} = 52.3$ V

Mark allocation: 2 marks

- 1 mark for correctly substituting values into the formula
- 1 mark for the correct answer

Note: It is possible to award consequential marks for using the value of the voltage obtained in

part f.: $V_{sec} = \dfrac{(\text{Answer } \textbf{part f.})}{3}$ V.

SET 5

Question 1a.

Worked solution

15 kW

The current in the transmission line is calculated from the power output of the Asterisk Powerbank, using $P = IV$.

Thus, $I = \dfrac{P}{V} = \dfrac{25 \times 10^3}{230} = 108.7$ A. The power loss in transmission is given by

$P_{loss} = I^2R = 108.7^2 \times 1.3 = 1.53 \times 10^4$ W.

Mark allocation: 3 marks

- 1 mark for correctly calculating the current in the transmission line
- 1 mark for correctly substituting values into the formula for power loss
- 1 mark for the correct answer in kW

Question 1b.i.

Worked solution

Stepping up the output voltage would result in a lower current in the transmission line for the same power delivery, according to conservation of energy. The lower current would then reduce the power loss, according to $P_{loss} = I^2R$

Mark allocation: 2 marks

- 1 mark for stating that stepping up the voltage would reduce the current in the transmission line
- 1 mark for relating the reduced current to a reduction in the power loss

Question 1b.ii.

Worked solution

Step-up transformers require AC power supply in order to work. This is because the alternating current produces an alternating magnetic flux, which induces an alternating current in the secondary windings.

OR

The direct current does not produce a changing magnetic flux, which is required to induce a current.

Mark allocation: 2 marks

- 1 mark for stating that step-up transformers require AC power **OR** for stating that DC does not produce a changing magnetic flux
- 1 mark for highlighting the requirement of an alternating magnetic flux to induce an alternating current

Question 1c.i.

Worked solution

$98\ A_{RMS}$

Since the inverter has an efficiency of 90%, using 25 kW of power input, the output of the inverter is $P_{out} = 0.9 \times 25 = 22.5$ kW. Since the output voltage is given in RMS, the output current may be calculated using $I = \dfrac{P}{V} = \dfrac{22.5 \times 10^3}{230} = 97.8\ A_{RMS}$. This is the current in the primary windings of the step-up transformer.

Mark allocation: 2 marks

- 1 mark for correctly calculating the power output of the inverter
- 1 mark for correctly calculating the output current of the inverter

Question 1c.ii.

Worked solution

124 W

Using the ideal transformer equation, $\dfrac{N_1}{N_2} = \dfrac{I_2}{I_1}$, the secondary current is given by $I_2 = \dfrac{I_1 N_1}{N_2} = \dfrac{97.8}{10} = 9.78\ A_{RMS}$.

Thus, when using the step-up transformer, the power loss is $P_{loss} = I^2 R = 9.78^2 \times 1.3 = 124$ W.

Mark allocation: 2 marks

- 1 mark for correctly calculating the secondary current of the transformer
- 1 mark for correctly calculating the power loss in the transmission line

Unit 4 | Area of Study 1   How has understanding about the physical world changed?

SET 1

Question 1

Answer: B

Explanatory notes

A passenger at rest relative to the spaceship will observe no length contraction. Instead, they will measure the 'proper length' of the spaceship.

Question 2

Answer: A

Explanatory notes

The extent of diffraction is proportional to $\frac{\lambda}{w}$. A smaller gap width will result in more diffraction.

Question 3

Answer: C

Explanatory notes

$v = f\lambda$

$\quad = 300 \times 0.03$

$\quad = 9.0 \text{ m s}^{-1}$

Question 4

Answer: D

Explanatory notes

Resonant frequencies, on a string fixed at both ends, have a frequency that is an integer multiple of the fundamental frequency $f_n = n f_0$, where $n = 1, 2, 3, \dots$.

SET 2

Question 1

Answer: **C**

Explanatory notes

Diffraction occurs when waves move through an aperture or bend around a corner.

Question 2

Answer: **D**

Explanatory notes

Since the wavelength remains the same, the wave equation $v = f\lambda$ is rearranged to obtain the wavelength of the 440 Hz sound in air: $\lambda = \dfrac{v}{f} = \dfrac{340}{440} = 0.7727$ m. The speed of sound in trimix is obtained by substituting the values into the wave equation: $v = 1300 \times 0.7727 = 1004.5$ m s^{-1}.

Question 3

Answer: **D**

Explanatory notes

The fringe spacing of the interference pattern is given by $\Delta x = \dfrac{\lambda L}{d}$.

Thus, the wavelength of the laser is

$$\lambda = \frac{\Delta x d}{L} = \frac{3.1 \times 10^{-3} \times 250 \times 10^{-6}}{1.912} = 4.05 \times 10^{-7} \text{ m} = 405 \text{ nm.}$$

Question 4

Answer: **C**

Explanatory notes

The energy, in joules, of the electrons that is needed to be converted is
$E_k = qV = 1.6 \times 10^{-19} \times 36 = 5.76 \times 10^{-18}$ J.

The de Broglie wavelength of the electrons is

$$\lambda = \frac{h}{\sqrt{2m_e E_k}} = \frac{6.63 \times 10^{-34}}{\sqrt{2 \times 9.11 \times 10^{-31} \times 5.76 \times 10^{-18}}} = 2.05 \times 10^{-10} \text{ m.}$$

 TIP

» **The formula for the de Broglie wavelength can be derived from the kinetic energy of the body:** $E_k = \dfrac{1}{2}mv^2 \rightarrow v = \sqrt{\dfrac{2E_k}{m}}$. **Since** $\lambda = \dfrac{h}{p} = \dfrac{h}{mv}$,

substituting for v will yield $\lambda = \dfrac{h}{m\sqrt{\dfrac{2E_k}{m}}} = \dfrac{h}{\sqrt{2mE_k}}$.

SET 3

Question 1

Answer: **A**

Explanatory notes

As the electron speeds up from rest to nearly the speed of light, the length of the stationary ruler, from the electron's frame of reference, contracts from 1 m to nearly zero.

The observed length, L, is related to the proper length, L_0, by $L = \dfrac{L_0}{\gamma}$. Since relativistic effects are low when the speed of the electron is low, there is little change in L initially. As γ increases from 1 (when the speed of the electron is zero) towards infinity (when the speed of the electron is close to the speed of light), the observed length of the ruler will contract towards zero.

Question 2

Answer: **C**

Explanatory notes

The wavelength of the radio signal can be found from $c = f\lambda$; thus, $\lambda = \dfrac{c}{f} = \dfrac{3 \times 10^8}{88.8 \times 10^6} = 3.38$ m.

Question 3

Answer: **C**

Explanatory notes

The energy per photon is given by $E = hf = 6.63 \times 10^{-34} = 88.8 \times 10^6 = 5.887 \times 10^{-26}$ J.

The number of photons emitted can be found from $n = \dfrac{P}{E} = \dfrac{12.5 \times 10^3}{5.887 \times 10^{-26}} = 2.12 \times 10^{29}$.

SET 4

Question 1a.

Worked solution

$v = f\lambda$

$340 = 500 \times \lambda$

$\lambda = \dfrac{340}{500}$

$= 0.68$ m

Mark allocation: 1 mark

- 1 mark for the correct answer

Question 1b.

Worked solution

	Perception of sound
At L	Louder than normal (constructive interference)
From L to M	Alternating loud and soft (nodes and antinodes)
At M	Quieter than normal (path difference = $\dfrac{3.8 - 1.14}{0.76}$ = 3.5 λ, destructive interference)

Mark allocation: 3 marks

- 1 mark for each correct answer listed in the table (up to 3 marks)

Question 1c.

Worked solution

Bess hears the 1000 Hz tone more loudly than the 11 000 Hz tone.

The lower frequency tone has a larger wavelength than the higher frequency tone.

The larger wavelength means the 1000 Hz tone undergoes more diffraction.

Mark allocation: 3 marks

- 1 mark for stating that the 1000 Hz tone is louder
- 1 mark for comparing the wavelengths
- 1 mark for mentioning that different levels of diffraction are caused by the wavelengths

Question 2a.

Worked solution

increase

Mark allocation: 1 mark

- 1 mark for correctly identifying the increase in the fringe spacing

Note: Diffraction effect (and the fringe spacing) increases with the increasing wavelength of the laser used.

Question 2b.

Worked solution

decrease

Mark allocation: 1 mark

- 1 mark for correctly identifying the decrease in the fringe spacing

Note: The fringe spacing decreases as the slit separation is increased.

Question 2c.

Worked solution

interference

Mark allocation: 1 mark

- 1 mark for correctly identifying the pattern

Note: Each slit acts as a point source of waves. The waves from these sources constructively and destructively interfere with each other, resulting in the alternating bright and dark fringes, which is known as an interference pattern.

Question 2d.

Worked solution

The pattern is an interference pattern, which is evidence of wave behaviour.

At locations where waves interact constructively there are bright regions, and where they interact destructively there are dark regions.

Constructive interference occurs when the path difference between the screen and the sources is an integer number of wavelengths ($n\lambda$), whereas destructive interference occurs when this path difference is $(n-1)\lambda$.

Mark allocation: 3 marks

- 1 mark for identifying interference patterns as a wave phenomenon
- 1 mark for explaining how interference patterns arise from constructive and destructive interferences
- 1 mark for explaining how the path difference between the two slits and the screen causes the interferences

Question 3a.

Worked solution

Chris is at rest relative to the object being measured and so records the proper length of the train. Deb is moving relative to the train and so records a contracted length relative to Chris.

$$\gamma = \frac{1}{\sqrt{1 - \dfrac{v^2}{c^2}}} = \frac{1}{\sqrt{1 - 0.98^2}} = 5.03$$

$$l = \frac{l_0}{\gamma} = \frac{10}{5.03} = 2.0 \text{ m}$$

Mark allocation: 2 marks

- 1 mark for the correct Lorentz factor (γ)
- 1 mark for the correct answer

TIP

> »   Make sure you are consistently able to identify proper length and proper time correctly.

Question 3b.

Worked solution

Deb is at rest relative to the falling ball, so she records the proper time.

$$\gamma = \frac{1}{\sqrt{1 - \dfrac{v^2}{c^2}}} = \frac{1}{\sqrt{1 - \left(\dfrac{2.97}{3.00}\right)^2}} = 7.09$$

$$t = t_0 \gamma$$

$$4.0 = t_0 \times 7.09$$

$$t_0 = \frac{4.0}{7.09} = 0.56 \text{ s}$$

Mark allocation: 2 marks

- 1 mark for the correct Lorentz factor (γ)
- 1 mark for the correct answer

Question 3c.

Answer

neither

Worked solution

Einstein's first postulate states that the laws of physics are the same for all observers. This means that no law of physics can identify a state of absolute rest.

Mark allocation: 3 marks

- 1 mark for circling 'neither'
- 1 mark for correctly stating (or paraphrasing) Einstein's first postulate
- 1 mark for stating that absolute rest does not exist or cannot be identified

Question 3d.

Worked solution

$$E = 25 \times 10^6 \text{ eV}$$
$$= 25 \times 10^6 \times 1.6 \times 10^{-19}$$
$$= 4.0 \times 10^{-12} \text{ J}$$
$$E = \Delta mc^2$$
$$4.0 \times 10^{-12} = \Delta m \times 9.0 \times 10^{16}$$
$$\Delta m = 4.4 \times 10^{-29} \text{ kg}$$

No protons or neutrons are destroyed.

Mark allocation: 3 marks

- 1 mark for the correct conversion to joules
- 1 mark for the correct value for mass
- 1 mark for stating that no particles are destroyed

Note: Some protons are converted to neutrons in this process.

SET 5

Question 1

Answer: **A**

Explanatory notes

Their relative motion to each other would cause both Gini and Jordan to measure length contraction in the other person's frame of reference; hence, both would observe the other person's ruler to be contracted by the Lorentz factor of 2.00.

> **TIP**
>
> » Length contraction is measured by observers in both reference frames that are moving relative to each other, and this can contribute to some confusion. For example, consider two reference frames, A and B, that are moving relative to each other. From the perspective of an observer at A, the length of an object at B that is measured by an observer at B is the proper length; whereas the observer at A would measure the object at B to have a shorter length, the apparent length. Likewise, an observer at B would also measure length contraction for objects at A.

Question 2

Answer: **B**

Explanatory notes

The stopping voltage is increased until even the most energetic electrons are repelled and the photocurrent drops to zero. This allows the experimenter to calculate the maximum kinetic energy of the ejected electrons.

Question 3

Answer: **D**

Explanatory notes

All the experiments described in options A, B and C demonstrate the interactions of light and electrons, which are best explained using the particle model.

Question 4

Answer: **C**

Explanatory notes

If a higher frequency light source is used, it would increase the rate of photoelectrons being liberated from the metal; thus, the magnitude of the photocurrent would increase.

SET 6

Question 1

Answer: **C**

Explanatory notes

Diffraction is a wave phenomenon, with the fringe spacing determined by the wavelength. Thus, this experiment demonstrates that the electron beam behaves in a wave-like manner despite being ordinarily thought of as a particle-like stream of matter. The similar fringe spacing is due to the fact that the electrons in the beam have a de Broglie wavelength that is similar to the wavelength of the X-ray.

Question 2

Answer: **C**

Explanatory notes

The energy associated with that photon of light is given by the equation

$$E = \frac{hc}{\lambda} = \frac{4.14 \times 10^{-15} \times 3 \times 10^8}{560 \times 10^{-9}} = 2.218 \text{ eV.}$$

TIP

» Since the unit of energy used in the question is the electron volt, using the electron volt second (eV s) version of Planck's constant saves time by removing the extra step of converting the answer from joules to electron volts. It is also important, when deciding which version to use, to check the units of the quantities involved in formulas to ensure consistency.

Question 3

Answer: **D**

Explanatory notes

The discrete nature of the harmonics of standing waves – that is, $n = 1, 2, 3, \ldots$ – helps explain how electrons orbit as matter waves at discrete energy levels. When electrons transition between energy levels, energy in the form of light is emitted, with the difference between the energy levels corresponding to the light frequency according to the equation $\Delta E = hf_{emitted}$.

Question 4

Answer: **C**

Explanatory notes

Diffraction is a wave phenomenon, so the electrons behave in a wave-like manner when passing through the salt crystal. The electrons behave in a particle-like manner when they hit the phosphor screen.

SET 7

Question 1

Worked solution

$E = hf$

$\quad = 4.14 \times 10^{-15} \times 9.42 \times 10^{14}$

$\quad = 3.90 \text{ eV}$

For the transition $n = 4$ to $n = 2$:

$\Delta E = 5.5 - 1.6$

$\quad = 3.9 \text{ eV}$

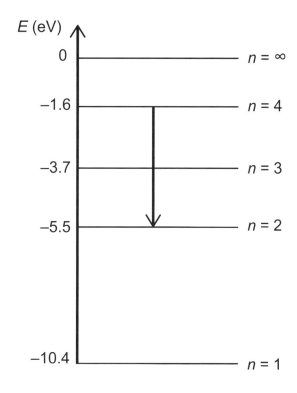

Mark allocation: 3 marks

- 1 mark for the correct energy
- 1 mark for an arrow between $n = 4$ and $n = 2$
- 1 mark if the arrow is downwards (between any energy levels)

 TIP

» **Emitted photons are always the result of a transition towards the ground state (a downwards arrow).**

Question 2a.

Worked solution

The work done by the electron gun is equal to the kinetic energy of the electron. Once the velocity of the electron is found, the wavelength can be calculated.

$$\frac{1}{2}mv^2 = qV$$

$$\frac{1}{2} \times 9.11 \times 10^{-31} \times v^2 = 1.6 \times 10^{-19} \times 15$$

$$v = \sqrt{\frac{1.6 \times 10^{-19} \times 15}{\frac{1}{2} \times 9.11 \times 10^{-31}}} = 2.30 \times 10^6 \text{ m s}^{-1}$$

$$\lambda = \frac{h}{mv} = \frac{6.63 \times 10^{-34}}{9.11 \times 10^{-31} \times 2.30 \times 10^6} = 3.17 \times 10^{-10} \text{ m}$$

Mark allocation: 3 marks

- 1 mark for the correct velocity
- 1 mark for substituting correctly into the de Broglie wavelength formula
- 1 mark for the correct answer

Question 2b.

Worked solution

The radiation must consist of photons that have the same wavelength as the electrons for the diffraction patterns to be identical.

$$c = f\lambda$$

$$f = \frac{c}{\lambda}$$

$$= \frac{3.0 \times 10^8}{3.2 \times 10^{-10}}$$

$$= 9.38 \times 10^{17} \text{ Hz}$$

Mark allocation: 2 marks

- 1 mark for substituting correctly
- 1 mark for the correct answer

Question 2c.

Worked solution

$p_{flea} = mv = 200 \times 10^{-6} \times 2.0 = 4.0 \times 10^{-4}$ kg m s^{-1}

$p_{violet} = \dfrac{h}{\lambda} = \dfrac{6.63 \times 10^{-34}}{400 \times 10^{-9}} = 1.66 \times 10^{-27}$ kg m s^{-1}

$\dfrac{4.0 \times 10^{-4}}{1.66 \times 10^{-27}} = 2.4 \times 10^{23}$ photons

Mark allocation: 3 marks

- 1 mark for the correct momentum of the flea
- 1 mark for the correct momentum of the violet photon
- 1 mark for the correct answer

SET 8

Question 1a.

Worked solution

Use the work function to calculate the threshold frequency:

$W = hf_0$

$2.9 = 4.14 \times 10^{-15} \times f_0$

$f_0 = \dfrac{2.9}{4.14 \times 10^{-15}}$

$\quad = 7.0 \times 10^{14}$ Hz

Mark allocation: 2 marks

- 1 mark for attempting to use the work function
- 1 mark for the correct answer

Question 1b.

Worked solution

Draw a trend line for the data.

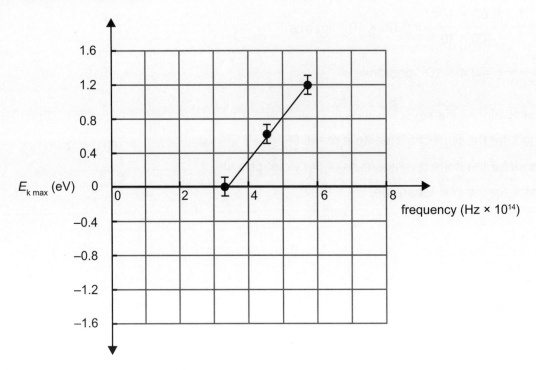

Calculate the gradient of the trend line:

$$h = \frac{1.2 - 0}{(5.7 - 3.2) \times 10^{14}} = 4.8 \times 10^{-15} \text{ eV s}$$

Mark allocation: 3 marks

- 1 mark for drawing the trend line
- 1 mark for attempting to find the gradient
- 1 mark for the correct gradient (accept values from 4.7×10^{-15} eV s to 4.9×10^{-15} eV s)

Question 1c.

Worked solution

Use the trend line to predict the $E_{k\,max}$.

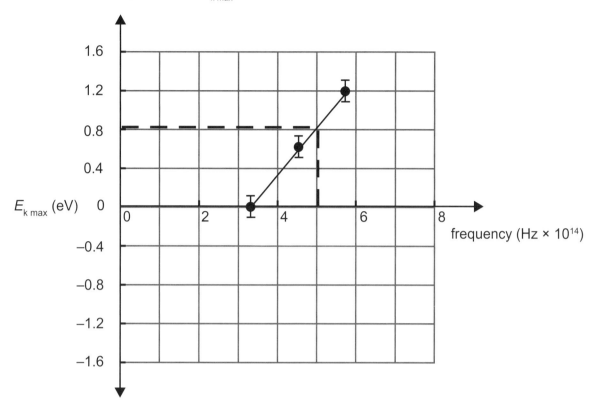

Convert from eV to V.

$E_{k\,max}$ = 0.81 eV indicates V_0 = 0.81 volts.

Mark allocation: 1 mark

1 mark for indicating the use of the trend line to predict the value of $E_{k\,max}$ (accept values from 0.70 V to 0.90 V)

Question 2a.

Worked solution

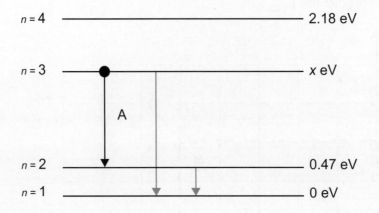

$n = 4$ ——————————— 2.18 eV

$n = 3$ ——————●——————— x eV

A

$n = 2$ ——————▼——————— 0.47 eV

$n = 1$ ——————————— 0 eV

Mark allocation: 2 marks

- 1 mark for each correct arrow (up to 2 marks)

Note: If more than two arrows are provided, remove a mark for each incorrect arrow (minimum 0 marks); e.g. two correct arrows and one incorrect arrow, award 1 mark.

Question 2b.

Worked solution

The energy of the emitted photon can be obtained from its momentum via the equation
$E = pc \Rightarrow 7.1 \times 10^{-28} \times 3 \times 10^8 = 2.13 \times 10^{-19}$ J.

This is equivalent, in electron volts, to $\dfrac{2.13 \times 10^{-19}}{1.6 \times 10^{-19}} = 1.33$ eV.

The energy of the photon is the difference in the energy levels between $n = 3$ and $n = 2$; that is, $x - 0.47 = 1.33 \Rightarrow x = 1.8$ eV.

Alternatively, the wavelength of the photon may be calculated from

$p = \dfrac{h}{\lambda} \Rightarrow \lambda = \dfrac{6.63 \times 10^{-34}}{7.1 \times 10^{-28}} = 9.34 \times 10^{-7}$ m, and the energy of the photon is then

$E = \dfrac{hc}{\lambda} = \dfrac{6.63 \times 10^{-34} \times 3 \times 10^8}{9.34 \times 10^{-7}} = 2.13 \times 10^{-19}$ J, as before.

Mark allocation: 3 marks

- 1 mark for obtaining the energy of the photon using the momentum
- 1 mark for equating the energy of the photon with the difference in the energy levels
- 1 mark for the correct answer

Question 2c.

Worked solution

The electrons in atoms such as those in element Z can absorb only quanta of energy that correspond to the difference in energy between any two levels. Since the energy of the beam of photons does not match any of the quanta that could be absorbed by element Z, the photons will pass through the sample without transferring energy to the electrons.

Mark allocation: 2 marks

- 1 mark for stating that the electrons can absorb only quanta of energy that correspond to differences between energy levels
- 1 mark for highlighting that the photon energy does not correspond to any quanta that could be absorbed

SET 9

Question 1

Worked solution

Under the wave model:

- The kinetic energy of ejected electrons should be proportional to the intensity of incident light.
- Electrons should be ejected from the metal by light at any frequency above a minimum intensity.
- A time delay should be observed, particularly at low intensities, as energy is gradually transferred to the metal by the light waves.

Mark allocation: 3 marks

- 1 mark for relating kinetic energy to light intensity
- 1 mark for mentioning that any frequency of light should eject electrons
- 1 mark for mentioning a time delay

 TIP

» **These questions may be answered in dot points. You may find this a quicker and easier way to express your ideas.**

Question 2a.

Worked solution

The stopping voltage gives the energy of the photoelectrons, $E = Vq$; thus, substituting the values gives $1.47 \times 1.6 \times 10^{-19} = 2.35 \times 10^{-19}$ J.

Mark allocation: 1 mark

- 1 mark for the correct answer

Question 2b.

Worked solution

Use $E_{electron} = E_{photon} - W$, where $E_{photon} = hf \Rightarrow 4.14 \times 10^{-15} \times 6.45 \times 10^{14} = 2.67$ eV.

The energy of the photoelectron is provided by the stopping voltage data in the question: 1.47 eV.

Rearranging the equation above gives $W = E_{photon} - E_{electron} \Rightarrow 2.67 - 1.47 = 1.2$ eV.

Mark allocation: 3 marks

- 1 mark for correctly calculating the energy of each photon
- 1 mark for correctly substituting values into the equation relating the photon energy, photoelectron energy and work function
- 1 mark for the correct answer to the work function

TIPS

» It is important to distinguish between the photoelectron energy (i.e. the kinetic energy of the photoelectron ejected from the metal) and the photon energy (i.e. the energy of a light photon, the 'input' energy). One way to check that you have correctly identified each is that the photon energy is always higher than the photoelectron energy.

» Also, by noting that the photoelectron energy is given in electron volts, using the electron volt second (eV s) version of Planck's constant could save some time by removing the need to convert from joules to electron volts.

Question 2c.

Worked solution

decrease

Mark allocation: 1 mark

- 1 mark for correctly identifying the decrease in the stopping voltage

Note: The light is now of a lower frequency, hence photons have less energy. The stopping voltage will now be less than before.

Question 2d.

Worked solution

stay the same

Mark allocation: 1 mark

• 1 mark for the correct answer

Note: The intensity of the light will only affect the magnitude of the current of the photoelectrons (provided the light frequency is above the threshold frequency), not their energy. Thus the stopping voltage will remain the same.

Question 3a.

Worked solution

Use the equation $E = \dfrac{hc}{\lambda} = \dfrac{6.63 \times 10^{-34} \times 3 \times 10^8}{2.0 \times 10^{-10}} = 9.95 \times 10^{-16}$ J.

Mark allocation: 2 marks

• 1 mark for correctly substituting values into the correct equation

• 1 mark for the correct answer

Question 3b.

Worked solution

From the de Broglie wavelength, we can obtain the velocity of the electrons by rearranging the equation:

$$\lambda = \frac{h}{p} = \frac{h}{mv} \Rightarrow v = \frac{h}{m\lambda} = \frac{6.63 \times 10^{-34}}{9.1 \times 10^{-31} \times 2.0 \times 10^{-10}} = 3.643 \times 10^6 \text{ m s}^{-1}.$$

The kinetic energy of the electrons is obtained from

$$E_k = \frac{1}{2}mv^2 = \frac{1}{2} \times 9.1 \times 10^{-31} \times (3.643 \times 10^6)^2 = 6.04 \times 10^{-18} \text{ J}.$$

Mark allocation: 3 marks

• 1 mark for correctly substituting values into the correct equation to obtain the velocity of the electrons

• 1 mark for the correct answer to the velocity of the electrons

• 1 mark for the correct answer for the kinetic energy

 TIP

» **For questions involving light and matter waves, remember that when the wavelengths of both the light waves (such as ultraviolet, X-rays or gamma rays) and the matter waves are the same, their momentum should also be the same. However, the energy of the light waves is higher than that of the matter waves, as shown in this question.**

SET 10

Question 1a.

Worked solution

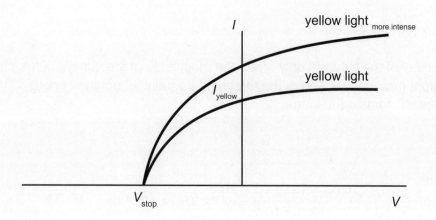

Explanatory notes

Using the same light frequency will produce photoelectrons with the same maximum kinetic energy; hence, the stopping voltage, V_{stop}, will not change. However, the higher light intensity will increase the number of photons incident on the metal plate, thereby increasing the number of photoelectrons and producing a higher photocurrent.

Mark allocation: 1 mark

- 1 mark for stating the same V_{stop} but higher photocurrent

Question 1b.

Worked solution

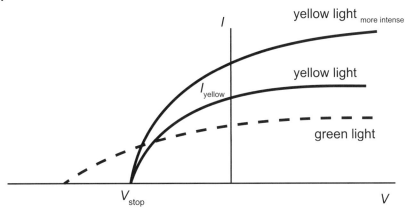

Explanatory notes

Green light has a higher frequency and therefore will produce photoelectrons with a higher maximum kinetic energy. Hence, the stopping voltage will have a greater magnitude; that is, it will be on the *x*-axis to the left of V_{stop}.

For the same light power output incident on the metal plate, a higher frequency light will have a correspondingly lower number of photons, according to the light beam energy equation $E_{beam} = nhf$. This will result in a lower photocurrent.

Mark allocation: 2 marks

- 2 marks for sketching the *x*-intercept to the left of the V_{stop} but having lower photocurrent

Question 2a.

Worked solution

Photons are emitted with a wavelength that corresponds to the difference of the energy levels that the electrons transition between.

The relationship $E_{difference} = \dfrac{hc}{\lambda} = \dfrac{4.14 \times 10^{-15} \times 3 \times 10^{8}}{627 \times 10^{-9}} = 1.98$ eV, as required.

Mark allocation: 2 marks

- 1 mark for selecting the correct equation and substituting the correct values
- 1 mark for obtaining the correct value of photonic energy

Question 2b.

Worked solution

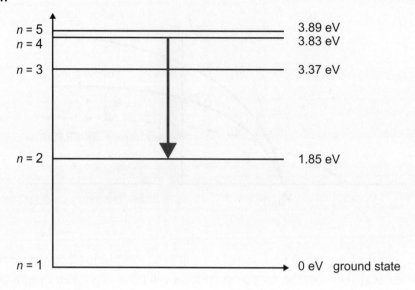

Explanatory notes

The electron transition is between $n = 4$ and $n = 2$; that is: $E_{difference} = 3.83 - 1.85 = 1.98$ eV.

Mark allocation: 1 mark

- 1 mark for an arrow between $n = 4$ and $n = 2$

Note: The tail of the arrow must touch only $n = 4$ and the arrowhead must touch only $n = 2$.

SET 11

Question 1a.

Worked solution

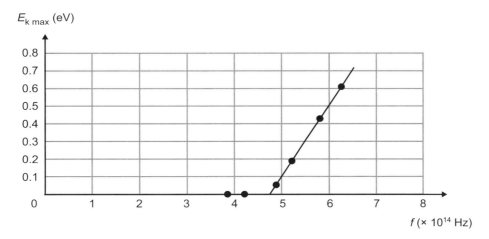

Mark allocation: 4 marks

- 2 marks for correctly plotted points

  Note: Deduct 1 mark for up to two errors and deduct 2 marks for three or more errors.

- 1 mark for correct axes scales

- 1 mark for the correct line of best fit, which must intersect the x-axis between 4.6×10^{14} Hz and 4.8×10^{14} Hz. Note that the first two values are below the threshold frequency and should not be included when determining the line of best fit.

Question 1b.

Worked solution

4.7×10^{14} Hz

Mark allocation: 1 mark

- 1 mark for the correct answer (accept 4.6×10^{14} Hz to 4.8×10^{14} Hz)

Note: It is possible to award consequential marks for using the x-axis intercept of the graph. Check for correct procedure and technique.

Question 1c.

Worked solution

1.9 eV

The work function of the metal is given by $W = hf_0$, where h is the Planck's constant obtained from the gradient of the graph, and f_0 is the threshold frequency obtained from the x-intercept of the graph. Thus, $W = 4.1 \times 10^{-15} \times 4.7 \times 10^{14} = 1.93$ eV.

Mark allocation: 2 marks

- 1 mark for correctly substituting the two values obtained in the experiment
- 1 mark for the correct answer

Note: It is possible to award consequential marks for using the answer from **part b.**; that is, the x-axis intercept of the graph, $W = 4.1 \times 10^{-15} \times$ (Answer **part b.**).

 TIP

» Be careful to read the full instructions in the question and use all information provided, such as the gradient of the graph. Often, students use the data given in the VCAA formula sheet, such as $h = 4.14 \times 10^{-15}$ eV s, rather than following instructions, losing valuable marks.

Question 2

Worked solution

3.4×10^{-16} J

Since the pattern produced by the electrons is similar to that produced by the X-rays, the de Broglie wavelength (and thus the momentum of the electrons) is the same as for the X-rays. The kinetic energy of the electrons may be calculated using

$$E_k = \frac{1}{2} m_e v^2 = \frac{1}{2} \frac{m_e^2 v^2}{m_e} = \frac{p^2}{2m_e}$$

Substituting in the values: $E_k = \dfrac{(2.47 \times 10^{-23})^2}{2 \times 9.1 \times 10^{-31}} = 3.36 \times 10^{-16}$ J.

Mark allocation: 3 marks

- 1 mark for deriving the kinetic energy equation from the momentum of the electrons

  Note: This mark may be awarded if the equation is simply copied from a personal formula sheet rather than derived.

- 1 mark for correctly substituting values into the equation
- 1 mark for the correct answer

Question 3

Worked solution

Emitted photons with a frequency of $f = 3.97 \times 10^{14}$ Hz are produced when electrons transition between energy levels with difference of $\Delta E = hf = 4.14 \times 10^{-15} \times 3.97 \times 10^{14} = 1.64$ eV. This does not correspond to any of the energy level differences for this element, which are $\Delta E_{3 \to 1} = 3.37$ eV, $\Delta E_{2 \to 1} = 1.85$ eV and $\Delta E_{3 \to 2} = 3.37 - 1.85 = 1.52$ eV.

Mark allocation: 3 marks

- 1 mark for the correct calculation of the photon energy of 1.64 eV

- 1 mark for the correct calculation of the energy level differences

- 1 mark for pointing out that the photon energy does not correspond to any of the energy level differences for this element

TIP

» It is vital to remember that problems involving electron transitions between energy levels require you to find the difference(s) between energy levels. Transitions from higher energy levels back to ground state ($n = 1$) emit photons with energies corresponding to the energy levels themselves; however, transitions between energy levels emit photons with energies corresponding to the difference between energy levels.

Unit 4 | Area of Study 2   How is scientific inquiry used to investigate fields, motion or light?

SET 1

Question 1

Answer: B

Explanatory notes

A discrete variable is measured in categories, as opposed to a continuous variable.

Independent variables are deliberately manipulated by the experimenter.

Question 2

Answer: B

Explanatory notes

A controlled variable is kept the same in all tests.

Question 3

Answer: D

Explanatory notes

A dependent variable is measured by the experimenter. An experiment is designed to investigate the effect that the independent variable has on the dependent variable.

Question 4

Answer: C

Explanatory notes

The absence of data points in the region of 0.5 m weakens the assumption of a linear relationship here.

Option A is incorrect because one may extrapolate a graph beyond the range of experimental data, based on the observed trend and the strength of the correlation between variables.

Option B is incorrect because actual measurements are not required, provided a causal link and a relationship could be modelled between the variables.

Option D demonstrates a reliance on a textbook rather than experimental evidence.

Question 5

Answer: B

Explanatory notes

Systematic errors are due to faults in the apparatus and equipment, and cannot be reduced by making repeated measurements and taking the average value of all measurements. All the other errors could be reduced or eliminated by having different experimenters and multiple measurements, which would tend to even out the errors.

SET 2

Question 1

Answer: **C**

Explanatory notes

The uncertainty for this set of three readings could be estimated as the difference between the average and the most extreme reading (that is, the reading furthest from the average). The

average of the readings is 441.467 (calculated from $\bar{x} = \dfrac{442.4 + 441.1 + 440.9}{3}$) and the

difference between the average and the most extreme reading is 442.4 − 441.5 = 0.9.

Since the individual readings are precise to 0.1 Hz, the average and uncertainty of these readings should be stated to the same number of decimal places.

Question 2

Answer: **C**

Explanatory notes

A hypothesis is still useful even though experimental evidence showed that it could not be supported, as the lack of support for a hypothesis through evidence gained can lead to new ideas and theories being developed to explain an observed phenomenon.

Question 3

Answer: **C**

Explanatory notes

The falling mass is constant; hence, the net force is constant, meaning that it is the controlled variable. The mass of the cart is manipulated, and the acceleration of the cart is measured. This leads to the conclusion that the mass of the cart is the independent variable (because it is manipulated), and the acceleration is the dependent variable (because its measured value depends on the mass of the cart).

Question 4

Answer: **A**

Explanatory notes

Since the force on the conductor changes according to the amount of current, it is a dependent variable because its magnitude depends on the amount of current. The amount of current is the independent variable, as its magnitude can take any value within the limits of the available apparatus and equipment. All other variables, including the length of the conductor and the magnetic field strength, must be controlled.

 TIP

» **Remember that any variable that is measured must depend on changes in the variable that is modified. Hence, measured variables are dependent variables, whereas modified variables are independent variables. All others must remain the same or be controlled.**

Question 5

*Answer: **A***

Explanatory notes

A constant friction force is a systematic error, which will affect only the accuracy of the measurement made with the onboard accelerometer.

SET 3

Question 1a.

Worked solution

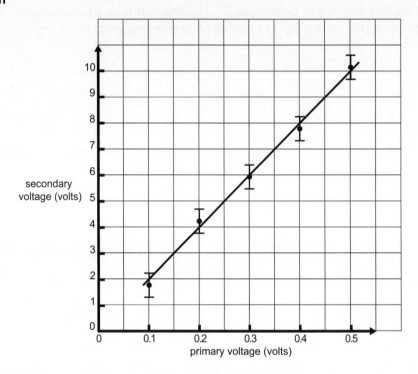

Mark allocation: 5 marks

- 1 mark for the correct scale and label on the *x*-axis
- 1 mark for the correct scale and label on the *y*-axis
- 1 mark for correctly plotted points
- 1 mark for correct uncertainty bars
- 1 mark for a correct trend line

Note: Subtract 1 mark if the axes are reversed (i.e. primary on the vertical axis, secondary on the horizontal axis).

Question 1b.

Answer

Highest possible value: 23.25

Lowest possible value: 18.5

Worked solution

The gradient of the line of best fit is the variable of the vertical axis (the DV) divided by the variable of the horizontal axis (the IV).

Use the error bars on the first and last point to draw lines with the maximum and minimum gradient, as shown below. Calculate the gradient of each line.

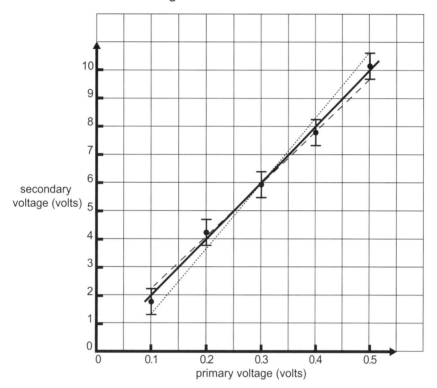

Maximum gradient:

$$\frac{10.5 - 1.2}{0.5 - 0.1} = 23.25$$

Minimum gradient:

$$\frac{9.6 - 2.2}{0.5 - 0.1} = 18.5$$

Mark allocation: 3 marks

- 1 mark for an attempt to calculate the gradient of the trend lines
- 1 mark for the correct maximum gradient (accept 22.75 to 23.75)
- 1 mark for the correct minimum gradient (accept 18 to 19)

Question 2a.

Worked solution

The dependent variable is the time taken to hit the bottom, or t.

The independent variable is the depth of oil, or x.

Mark allocation: 2 marks

- 1 mark for the correct dependent variable
- 1 mark for the correct independent variable

> **TIP**
>
> » The dependent variable is the variable that responds to changes in the independent variable. Since the depth of oil is changed, while the time taken to hit the bottom responds to the changes in the depth, their roles as independent or dependent variables can be distinguished.

Question 2b.

Worked solution

There are many possible answers.

It may be a physical characteristic of the oil: the oil temperature, type of oil or the viscosity of the oil.

It may be a physical characteristic of the ball: ball mass, ball diameter, surface roughness of ball or the ball density.

Mark allocation: 1 mark

- 1 mark for a correct example

Question 2c.

Worked solution

When repeat measurements are made, it is expected that each measurement will cluster about the true value, provided there is no systematic error involved.

Repeat measurements also give insight into the random error associated with a measurement by providing an estimate of the uncertainty of the measurement.

Mark allocation: 1 mark

- 1 mark for an acceptable explanation for why repeat measurements should be made

Question 2d.

Worked solution

The dimension of the uncertainty bar for that data point is approximately 0.8 s, so the uncertainty is ± 0.4 s. However, a range of answers between 0.30 s and 0.45 s is acceptable. 0.5 s is not acceptable as the uncertainty bar is clearly less than 1 s.

Mark allocation: 1 mark

- 1 mark for an answer between 0.30 s and 0.45 s

Question 2e.

Worked solution

Time taken to fall is proportional to the square root of the depth of oil: $t \propto \sqrt{x} \Rightarrow t = k\sqrt{x}$.

The data look best when modelled as $t \propto \sqrt{x}$. The first option is an inverse relationship, where t reduces towards zero as x is increased. The second option is a parabola, where t increases exponentially as x is increased.

Mark allocation: 1 mark

- 1 mark for the correct choice

Question 2f.

Worked solution

Using the third model, $z = \sqrt{x}$:

z, new variable to test the model chosen
0.32
0.45
0.89
1.10
1.41
1.73

Mark allocation: 2 marks

- 1 mark for three to five correct calculations
- 1 additional mark if all six calculations are correct

Note: Consideration should be given to correct calculations of the wrong model chosen from **part e.** (consequential marks).

Question 2g.

Worked solution

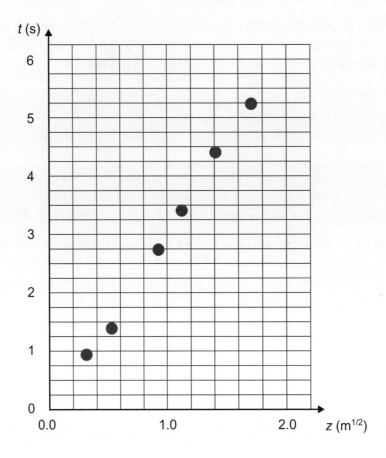

Mark allocation: 2 marks

- 1 mark for the correct scale and labelling of the *x*-axis

- 1 mark for correctly plotting the data

Note: Consideration should be given to correctly plotting the wrong model chosen in **part e.** (consequential marks).

Question 2h.

Worked solution

Since the graph appears to be linear, it supports the hypothesis that the time, *t*, varies proportionally with the square root of *x*.

Mark allocation: 1 mark

- 1 mark for an answer that supports, or refutes, the model chosen in **part e.**

SET 4

Question 1a.

Worked solution

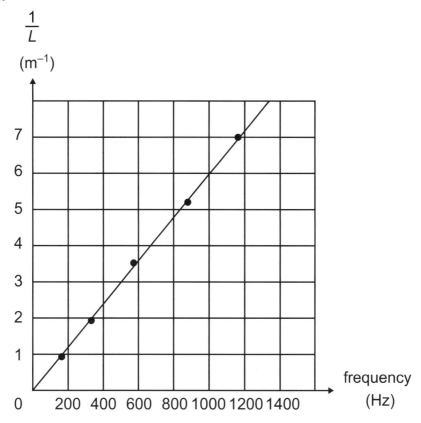

Mark allocation: 5 marks

- 2 marks for correct scales and units on each axis
- 2 marks for correctly plotted points (deduct 1 mark for up to two errors; deduct 2 marks for three errors or more)
- 1 mark for the correct line of best fit

Question 1b.

Worked solution

0.006 s m^{-1}

This value is obtained by taking any two points on the line of best fit and substituting their coordinates into the formula for the gradient of a straight line.

Mark allocation: 2 marks

- 1 mark for selecting two points that are on the line of best fit and correctly substituting their coordinates into the formula for gradient
- 1 mark for a gradient between 0.0057 and 0.0063

Note: It is possible to award consequential marks for using a different line of best fit if the correct procedure and calculation are used.

TIP

» You must use the line of best fit that you drew to calculate the gradient. Do not calculate the gradient from two actual data points in the table. When possible, use two data points on the line of best fit that are at the intersection of grid lines (e.g. (0, 0) and (1000, 6)), which give the value in the solution $m = \dfrac{6 - 0}{1000 - 0} = 0.006$. Otherwise, use two data points on the line of best fit where it crosses a horizontal grid line (e.g. (180, 1) and (680, 4)), which give a value equal to the one in the solution: $m = \dfrac{4 - 1}{680 - 180} = 0.006$.

Question 1c.

Worked solution

333 m s^{-1}

Gradient $m = \dfrac{L^{-1}}{f} = \dfrac{1}{fL}$ and $L = \dfrac{\lambda}{2}$; therefore, $m = \dfrac{2}{f\lambda}$.

So $v = f\lambda = \dfrac{2}{m} = \dfrac{2}{0.006} = 333 \text{ m s}^{-1}$.

Mark allocation: 3 marks

- 1 mark for correctly deriving the equation relating v and m
- 1 mark for correctly substituting the gradient into the equation
- 1 mark for correctly calculating a value between 317 m s^{-1} and 351 m s^{-1}

Note: It is possible to award consequential marks for using a different gradient if the correct procedure and calculation are used.

Question 2a.

Worked solution

systematic error

Mark allocation: 1 mark

• 1 mark for correctly identifying the type of error

Question 2b.i.

Worked solution

accuracy only

Mark allocation: 1 mark

• 1 mark for correctly identifying that the error will only affect accuracy

Question 2b.ii.

Worked solution

A systematic error results in a deviation away from the true value and, therefore, reduces accuracy. It does not affect precision, which is how close multiple measurements are to each other.

Mark allocation: 2 marks

• 1 mark for demonstrating an understanding that systematic errors affect accuracy
• 1 mark for demonstrating an understanding that precision is unaffected by systematic errors

● Acknowledgements

Insight Publications thanks Francis Dillon and Nick Howes for writing and reviewing this resource, and Moses Khor for substantial writing contributions.